GUIDE
TO
ASTRONOMY

GUIDE
TO
ASTRONOMY

David Geddes

CAXTON REFERENCE

© 2001 Caxton Editions

This edition published 2001 by Caxton Publishing Group Ltd,
20 Bloomsbury Street, London, WC1B 3QA.

Design and compilation by The Partnership Publishing Solutions Ltd,
Glasgow, G77 5UN

Printed and bound in India

CONTENTS

Introduction

Astronomy is one of the oldest sciences and one which has intrigued and fascinated mankind since the beginning of time. It is often confused with astrology, much to the chagrin of astronomers and amateurs deeply interested in the subject. One of the most popular astronomers in the country has said that astrology, which suggests that human character and destiny are influenced by the stars, is totally without foundation, and the best that can be said for it is that it is fairly harmless so long as it is confined to seaside piers, circus tents and the columns of tabloid newspapers. Now you know!

At 10.56 p.m. Washington time on 20 July 1969, 3.56 a.m. on 21 July in London, a huge, incredulous, worldwide audience first listened and then watched as the first man stepped onto the moon. On the radio we were able to follow every manoeuvre of the lunar module Eagle as it neared the surface of the moon and we listened with bated breath to the conversations between the astronauts and Mission Control at Houston in Texas. At last the transmission came from Mission Control, 'We copy you down, Eagle.' and Neil Armstrong's reply,

'Houston, Tranquillity Base here, the Eagle has landed.'

Mission Control confirmed, 'Roger, Tranquillity, we copy you on the ground. You've got a bunch of guys about to turn blue. We're breathing again. Thanks a lot.'

This writer followed every word and then went out to the garden to gaze in wonder at the moon, which was now inhabited by two men from earth. After a few hours, we were then able to watch as Neil Armstrong descended from the lunar module, placed his left foot on the moon and uttered his famous statement, 'That's one small step for man, one giant leap for mankind'.

The landing on the moon must have signalled the high tide in world interest in astronomy and space travel, but in a year or two that interest had waned and a flight to the moon seemed to become commonplace, although it had been the dream of man for thousands of years. The world's attention span is not very long! The only other period of high interest came when tragedy almost struck Apollo 13. Since these days there has been an increased awareness of astronomy, however, and television programmes by experts such as Patrick Moore have

appealed to a large audience. The great attraction of
astronomy is that amateurs can not only enjoy an
absorbing pastime and hobby but also, through their
observations, make contributions to the
advancement of knowledge. This is one science
where the professionals really do appreciate the help
of amateurs, who, over the years, have made many
important discoveries including comets, minor
planets and supernovae.

It is hoped that this little guide will encourage
more people to take a greater interest in the sky
above them. Even in these light polluted days,
much can be seen with the naked eye and a simple
telescope is not an expensive item. It is really
thrilling to observe the surface of the moon in detail
or, using your telescope, study the rings of Saturn,
the phases of Venus or discover Andromeda, which
is a galaxy as immense as our own. This is not a
cold science. The sky and the stars have always
captivated man and influenced his emotions as well
as his mind. R.L. Stevenson in *Travels With a Donkey*
said 'No one knows the stars who has not slept, as
the French happily put it, a la belle etoile. He may
know all their names and distances and magnitudes,
and yet be ignorant of what alone concerns
mankind, – their serene and gladsome influence on
the mind. The greater part of poetry is about the
stars; and very justly, for they are themselves the
most classical of poets. These same far-away worlds,

sprinkled like tapers or shaken together like a
diamond dust upon the sky…'.

CHAPTER 1
The Beginnings of Astronomy

China

To trace the beginnings of astronomy we go back to the very start of recorded history. It is difficult to get a really cohesive picture of the development of this very ancient science as we are dealing with an immense span of time, during which observations, investigations and the formulation of theories were being carried out in widely disparate parts of the world. Although astronomers are adamant that their discipline should not be confused with astrology, it is impossible to know which science came first. In the beginning most astronomers were also astrologers and the real division between the two did not come until the seventh century AD. The ancient peoples studied the skies with a view to divining what lay in the future and this portent astrology required detailed and regular observation of the skies. It was important that the results of these observations were carefully documented and this is, perhaps, the biggest contribution made by the Chinese to early astronomy and to it's future development.

The Chinese believed that their country was the centre of the world and, indeed, in Chinese the name of the country is 'Middle Kingdom'. Their belief was that the stars and planets came under the authority of the emperor. Over very many years Chinese astronomers observed and noted solar eclipses until they were able to establish a cycle in which the sun, moon and earth were aligned in the same way every 18 years or so. They, therefore, became quite accurate in their prediction of eclipses. During eclipses the Chinese reckoned that the sun was being attacked by a dragon, which seemed to take bites out of it and the way to combat this was to gather as many people as possible to make as much noise as possible, with a view to frightening off the dragon. This, of course, always worked and so the emperor's sun would be saved once again!

Chinese astronomers made very great contributions to the study of the science. Although coming from the unique Chinese position of the belief in the total involvement of the people with the emperor and with natural events, they produced calenders which worked, charts of the stars, sophisticated astronomical instruments and the development of clocks to control the instruments. The Chinese also contributed theories to cosmology, the science or study of the universe. They proposed three main possibilities. The first was the theory of the sky as a gently curved roof above and parallel to the earth. They estimated that the distance between the earth and the sky was 46,000 km. The second theory was developed about 100 years BC and is referred to as the 'celestial sphere' or 'enveloping sky' theory. It described the heavens as being like a hen's egg with the earth lying in the centre, like the yolk of the egg. The third Chinese cosmological theory was that the sun, moon, planets and stars all float freely in infinite space. This theory emerged in the second and third centuries AD, although it is said to reflect much earlier thinking. This 'infinite empty space' theory ran into trouble with later Chinese astronomers, but it is interesting to note that, in addition to the great contributions already mentioned, the Chinese were the first to propose the possibility of an infinite universe.

Egypt

The basis of what could be called astronomy in Egypt goes back into the mists of time before 3000 BC. It was from these times that the early study of the skies formulated into myths which eventually became the the core of Egyptian religion. It was also at this stage of the development of Egypt that the observation of the movements of the sun and the moon resulted in the creation of a time unit of 365 days and a fairly sophisticated lunar calendar. These were created for purely religious reasons, in order to set the timings for offerings and feast days, but gradually a simplified version of the calendar was introduced to enable the people to carry out business dealings and structure their normal lives.

In these earliest times, the sun god Ra emerged as the most important god and the annual movement of the sun was noted and its turning points in the north and south were called the solstices. The ancient Egyptians came to see the Milky Way as depicting the female god Nut, who conceived and gave birth to Ra at times which were recognised by observing the sky. The belief that Ra recreated himself though Nut established the matrilineal inheritance of the royal line in Egypt. The worship of Ra also explains the siting and the shape of many of the temples and especially the pyramids. The famous pyramids at Giza are thought to reflect the way in which clouds and blowing dust in the sunlight seemed to portray stairways leading to heaven. The pyramids, therefore, were built as stone stairways to the heavens, by which the soul of the dead pharaoh could reach the northern stars, which were known as 'the immortal ones'.

The Egyptians were well aware of the movements in the skies and these movememts governed their religious life and also their secular calendar. The experts of today, however, consider that the only really significant scientific advances which they left to us are the civil calendar of 365 days and the division of day and night into 12 hours each.

Greece

The Egyptians had a remarkable grasp of mathematics, witness the fact that they had measured the fall of the Nile over a distance of some 700 miles with an error of only a few inches. They had also discovered, and had used in practice, the fact that the square on the hypotenuse of a right-angled triangle is equal to the sum of the squares on the other two sides. The ancient Greeks did not think that way. Their thought had dealt with moral, religious and social problems and most speculation on the physical universe had pondered on how it had come into existence rather than how it worked. The first Greek to express his ideas on astronomy in logical instead of mythological terms was Thales of Miletus. He was a merchant who had travelled to Egypt and had learned something of mathematics in that country and also Chaldean astronomy. He knew enough to be able to predict that there would be a total eclipse of the sun during the year 585 BC, which did indeed take place.

Long before the time of Thales, the Greeks regulated their agricultural activity by observing the rising and setting of major stars or groups of stars, such as Sirius, Arcturus or Orion, the Pleiades and the Hyades. From these early beginnings came almanacs and calendars. As early as around 350 BC, the great philosopher and scientist Aristotle made a very significant advance. The assumption from the beginning of time was that the earth was flat, but Atristotle was not convinced. He realised from the curved shadow of the earth on the moon during an eclipse that it must be a sphere, not flat. He then went further to prove his theory by observing the southern star Canopus. He pointed out that the star was visible from Alexandria in Egypt but not further north in Athens. This would not be the case if the earth was flat. This theory had been proposed by Pythagoras some 200 years earlier but the proof provided by Aristotle, and the respect in which he was held, convinced the Greeks of the day.

Another important figure in Greek astronomy was Hipparchus, who lived around 150 BC. He was born in Asia Minor, although he carried out most of his work in Rhodes. He was conversant with the great volume of knowledge on astronomy which was held at Babylon, although it is not clear how he had gained access to this. It is known that astronomers in Mesopotamia had amassed observational records going back as far as the eighth century BC and, in particular, they had perfected accurate ways of calculating and predicting events affecting the moon and the planets. Hipparchus was undoubtedly a genius and he made great contributions in the areas of mathematical geography and the design of instruments. His major achievement, however, was to understand the vast amount of imformation which he obtained from Babylon, adapt it to a calendar recognised in Greece, and so spark off a revolution in Greek astronomy.

One of the most important characters in Greek astronomy, and perhaps the one most remembered today, was Claudius Ptolemaeus, better known as Ptolemy. Working in Alexandria between 140 and 180 AD, he improved and added to the work done by Hipparchus and produced a great volume, known as the Almagest. This book has come down to us from the Arab translation and is a synthesis of all the accumulated Greek astronomical knowledge from the earliest times. Ptolemy produced the first map of the world based upon astronomical observation and he developed a model of the solar system with the earth at the centre. This geocentric model had long been accepted in Greece and was generally approved until the heliocentric theory of Copernicus appeared in 1543. Ptolemy brought together the various strands of Greek astronomy and it has been said that true astronomy began with the Greeks.

Mesopotamia and India

We have seen how the wealth of astronomical knowledge in Mesopotamia was taken by Hipparchus and incorporated into the body of Greek learning on the subject. What is clear is that astronomy and astrology were inextricably linked in this area and it would appear that regular observation of the moon and the planets began in Babylon. Babylonia and Assyria were the main components of Mesopotamia. All forms of divination were practiced, the study of the skies being an important part of this facination with possible future events. If, for instance, it had been noted in the past that an eclipse of the moon had taken place at a certain date of a certain month and soon tbereafter a great king had died, then it would be assumed that there was a risk of a similar death when the next such eclipse occurred. The king and his court were the centre of all this activity and regular messages reached the king from astronomer-astrologers, who were constantly monitoring the movements of the sun, moon and planets in various parts of the country.

A calendar was in use from the third millennium BC and this was luni-solar, the month beginning on the evening when the lunar crescent was first visible. The months seem to have been of 30 days. In the old texts, reference is made to a water-clock, which may have been used initially to mark the three

watches of the night. The clock would have been
filled at the beginning of the watch, which would
end when the water-clock was empty. These
primitive clocks became more and more
sophisticated until they would be able to be used for
quite complicated astronomical calculations.
Mesopotamian cosmology was limited and the study
of the planets and stars was completely bound
with religion and the place of the gods in the
scheme of things. One of the main features of the
astronomy carried out in Mesopotamia was the
meticulous listing and detailing of information and
observations. This is what was incorporated into
Ptolemy's Almagest and had such a significant
influence on Greek astrology.

The knowledge of astronomy arrived in India
from Mesopotamia and in India also it was very
closely tied to religious rituals. The early texts
appeared probably just after 1000 BC and they gave
information and advice on the carrying out of the
great rituals, which had to be performed at set times
within the solar year. Between the second and
fourth centuries AD, revised information on the
planets began to appear in Sanskrit texts. These
texts were translations from Greek treatises which
had been based on theories formulated in Babylonia
as far back as the third century BC. The texts
contained new information on astronomy and also
on astrology, which was then adapted to conform to

the Indian culture. There were various schools of thought throughout India and some of the calculations made were amazingly accurate, which suggests that the Indian astronomers were in contact with international developments and were also capable of very accurate time measurements. The Indian astronomers seemed to take theories and observations from other cultures, check them thoroughly, and then adapt them for their major requirements, which were the computations of calendars, the prediction of solar and lunar eclipses, time-keeping and the casting of horoscopes. Observation of the skies was not a major feature of astronomy in India, but rather the use of mathematics. Many of these mathematical advances were not recognised outside of India but some did travel and influenced astronomy in the Islamic world and Western Europe.

In a relatively short time after the death of the Prophet Muhammed in 632 AD, the Arabs had established an area of Muslim influence which ranged from Spain to Central Asia and India. The new Islamic culture brought together the old knowledge of the heavens from Arabia and the Middle East and merged this with the mathematical traditions of the Indians, Persians and Greeks to establish its own individual style of astronomy. What could be called Islamic astronomy was in place by the tenth century AD. The Koran states that man should use the celestial bodies to guide him and the sun, moon and stars are mentioned in the holy book. Accurate star catalogues were drawn up, the movements of the sun, moon and planets were measured more precisely than before and the Arabs made a real contribution to the sum of the knowledge available on astronomy. Advances were made in the area of astronomical instruments and observatories were set up, including an elaborate one in Samarkand in the 15th century. From about this time, however, astronomy in the Islamic world declined and less and less innovative work was done.

Europe

During much of the period between the death of Ptolemy and the fall of Constantinople to the Turks in 1453, astronomy in Greece and Europe generally seemed to stagnate. It was still a well-respected science which received a great deal of study, but very few new avenues were opened or new thinking put forward. Ptolemy's belief that the earth was positioned at the centre of the universe was studied and discussed, but no new radical thinker came forward to question it.

This period coincided with the rise of Christianity and much of the learning took place in the new monastic schools, which concentrated on the understanding of the bible and gave scant attention to research and observation in relation to astronomy. One very important advance did, however, take place around 570 AD. The Christian belief in one god was gradually taking over from the many spirits and gods who had been thought to govern all things in nature and at this time Bishop Isidore of Seville drew a distinction between astronomy and astrology for the first time. In his *Twenty Books of Etymologies* he showed that everything concerned with nature and man could be considered and discussed without referring to mythology. This persuaded people that there was a real difference between astronomy and astrology and that the former was a serious science, whereas the latter was merely a superstition which depended on some connection between the twelve signs of the zodiac and the organs of the body. The Christian calendar required astronomical help to fix the times for movable feasts such as Easter and Lent and these were covered extensively in the major work written in 725 by the Venerable Bede of Jarrow – On the Theory of Time-reckoning. This was an impressive example of monastic science, and it is interesting to note that in another of his volumes Bede, for the first time, numbered the years from the birth of Christ.

Towards the end of the first millennium AD,

education was expanding through monastic and cathedral schools and astronomy was taught as one of the seven 'liberal arts'. Latin sources were the most used but in time had been completely covered, and the scholars were becoming aware of the vast amount of material from Greek sources, which the Muslim world had retrieved and were available in Arabic translations. Many of these students travelled to Spain to study this new seam of learning which was possessed by the Moors and many of them settled there to work side-by-side with Muslim and Jewish scholars. Others returned home, bringing their new knowledge and experience to their schools. This cultural exchange had a tremendous impact on the study of astronomy and science generally in Europe.

As Europe moved out of the Middle Ages towards the Renaissance, the greatest transformation in astronomy since the time of Ptolemy was taking place. This rebirth of astronomy is very firmly linked to four men, who became great names in the science. The first was a German mathematician and astronomer whose name was Johann Muller but who called himself Regiomontanus, being the Latinized form of Konigsberg, the place where he was born. He was a child prodigy and in 1447, when he was only 11 years old, he enrolled in the University of Leipzig. At that age he devised a daily ephemeris for the sun, moon and planets for the year 1448. An ephemeris is a special almanac giving the daily positions of celestial objects. In 1450 he moved to the University of Vienna, where he collaborated with an astronomer called Georg Peurbach on an abridged translation of Ptolemy's Almagest. Peurbach died during the work and asked his younger colleague to complete the book, which was published in 1463 as the *Epitome of the Almagest*. Regiomontanus was extremely critical of most of the astronomers of his day and he worked tirelessly to point out the faults which had entered the science since the establishment of the theories laid down by Ptolemy. Regiomontanus was very highly respected and his influence did a great deal to bring back the practice of scientific observation to astronomy. Pope Sixtus IV asked him to go to Rome to organise a revision of the Julian calendar, but Regiomontanus died in 1476, before the work was completed.

Only three years before the death of Regiomontanus, the next great figure to appear in European astronomy was born. This was Nicolaus Kopernik, known as Copernicus. He was born in what is now a part of Poland and took up a career in the church. He studied at the universities of Bologna, Padua and Ferrara and it was at Bologna that he worked with the professor of astronomy and made his first observations. It would appear that he gave less and less time to his administrative duties in the church and concentrated on astronomy.

Compared to Regiomontanus, Copernicus was not as brilliant, but he was careful and hard-working. He was meticulous in his work and, perhaps surprisingly, he formulated the single most famous discovery, or theory in the entire history of astronomy. During his studies of the Ptolemaic theories, he came to the conclusion that the central premise that the earth was at the centre of the solar system was wrong. If the sun was the centre of the system the movements of the planets made more sense. Copernicus was very aware that his sun-centred theories would be violently opposed by the church and he proceeded cautiously. This would be seen as heresy and could put him in danger. He published a short manuscript privately in 1514 and did not issue the complete theory until 1536, when he was immediately attacked many church leaders, including Martin Luther. Full publication did not take place until 1543, the year of his death.

The next important character in Renaissance astronomy was born only three years after the publication of the theories of Copernicus. He was a Danish nobleman called Tycho Brahe who, as a youth, studied law but then became facinated by astronomy and embarked on study of the science on his own. Tycho was no believer in the ideas of Copernicus, in fact he could not for a moment consider that the earth was not the centre of the universe. His great contribution was his remarkably

accurate observations and he had many successes.
King Frederick II of Denmark was his patron and
gave him land and money to build an observatory
on an island in the Baltic Sea. The observatory was a
massive and extraordinary building and it was here
that Tycho carried out most of his painstakingly
accurate observations. He compiled a star catalogue
which was the best to date and he made
measurements of the movements of the planets.
Tycho must have been a very hard man to work with
and, in fact, it is reckoned that over a period of 20
years he had as many as 60 assistants. He produced
many innovations in instrumentation and it is sad
that none of his instruments have survived, nor has
the observatory. He formulated a theory, which
became known as the Tychonic System, in which the
five planets travel round the sun while the sun and
moon move round the earth, which Tycho believed
did not move. It is thought that he fell out of favour
with King Frederick's successor, King Christian IV
and he left Denmark forever in 1597. In 1599
Tycho was appointed Imperial Mathematician of the
Holy Roman Empire and the Emperor gave him an
estate near Prague. A former assistant,
Longomontanus, came back to join him and then a
new assistant, Johannes Kepler, also joined him to
take over part of the work. Tycho was at this time
still driving himself, and his assistants, hard but he
was no longer designing new instruments or making

new observations. It was then, with his new work just beginning, that Tycho Brahe died unexpectedly in 1601. His work was to be carried on spectacularly by his new assistant, Johannes Kepler.

Kepler was the fourth great name in Renaissance astronomy and was born in Germany in 1571. From his time at the University of Tubingen he learned of the heliocentric theory of Copernicus and he was convinced of its truth from the beginning. Kepler could not have been more unlike Tycho. Tycho was arrogant, ruthless and flamboyant, whereas Kepler was neurotic and frail, having been a sickly child and he also suffered from bad eyesight. The two men argued constantly and yet they were a perfect team, with Tycho's penchant for observation and Kepler's ability as a brilliant theorist. Kepler had already done good work before he joined Tycho, but his greatest contributions came after the death of Tycho. He was now in the position to delve into the mass of data which Tycho had been gathering for years on the movements and positions of the planets and use this data to solve the problems which he could recognise in the sun-centred system of Copernicus. He had complete faith in the observations of Tycho and eventually, after years of study, he was able to solve the greatest problem. He concluded that the planets did indeed orbit the sun – but in elliptical orbits, not the perfect circular orbits as stated by Copernicus. Following this major discovery, Kepler then went ahead to draw up his three Laws of Planetary Motion, which have since been the starting point for all further work on the subject. The first law states that the planets move in an elliptical path, with the sun in position at one focus of the ellipse. The second law is a bit more complicated and harder to explain in a few words. It concerns the radius vector, which is an imaginary line which joins the centre of a planet to the centre of the sun and the fact that it sweeps out equal areas of the ellipse in equal times. This results in a planet moving fastest when closest to the sun and slowest when it is furthest away. Both of these laws were published in 1609 and the third did not appear until 1618. The third law is even more complex and links a planet's orbital period and its distance from the sun, making it possible to compile a scale model of the Solar System. Although Kepler, through outstanding mathematical ability and investigation, discovered that the orbits of the planets were elliptical rather than perfect circles, the ellipses travelled by most of the planets are very close to circles. The most erratic is Mars, which has a difference of over twenty million miles between its closest and its furthest points from the sun.

We have reached a point in the history of astronomy where we should find out if there was any interest in the subject in other areas of the world which we have not so far considered. We will briefly look at what was happening in these areas.

The Americas

North and South America are reckoned to have been
populated by races from Asia who moved into North
America by crossing a land bridge over the Bering
Sea from what is now the far east of Russia to
Alaska. This is thought to have happened over
10,000 years ago and the bridge is, of course, no
longer there. These peoples were completely
untouched by the advances in culture, science and
particularly astronomy in the rest of the world, and
yet they were looking at the same sun, moon,
planets and stars. It is facinating to know what
influence the skies had on the inhabitants of the
New World. Various cultures established
themselves, but unfortunately, many of them did not
develop writing, through which they could pass on
their knowledge to future generations. In addition
to this, the Spanish conquerors of Central America
destroyed most of the Maya and Aztec documents
which they found.

From the sixth century BC in Central America,
organised societies were measuring time and
detailing their findings in writing. There are stone
carvings depicting calendars at a ruin near the
modern city of Oaxaca and many inscriptions of a

later period at Monte Alban nearby. The Maya used observations of the skies to produce a calendar which was as accurate as those produced by the Chinese. Later the Aztecs used celestial observations also, and throughout Mexico can be found Mayan and Aztec buildings which appear to be aligned to mark sunrise on the days of the equinoxes, the beginning of spring and autumn. Further north in what would now be southern Illinois, a North American people who have become known as the Mound Builders raised over one hundred earthen mounds, which seem to illustrate some kind of a cosmic layout.

In Mexico and Peru, signs of the high cultures of the Maya, Aztecs and Incas can be detected in the remains of cities which have clearly been laid out in astronomical designs. These societies were obviously very well organised and must have had fairly sophisticated systems of positional astronomy based upon accurate horizon observations. It is, however, clear that the peoples of the Americas did not grapple with questions such as whether the sun or the earth was at the centre of the universe or whether the earth was round or flat. They did not have banks of information like the Babylonians or theories on space or orbits like the Greeks. This does not mean, however, that these native Americans did not wonder or theorise about the world they lived in. Their concerns were immediate and revolved around day-to-day life, the timing of the seasons for agriculture and religious rites.

Africa and Oceania

During the nineteenth century Africa was known to Europeans as the Dark Continent. It is surprising how little knowledge there was regarding any aspect of life on the continent and there certainly appeared to be no contact with outside cultures on matters astronomical. Apart from Egypt, there are no African written records before 1800 and little was known of traditional myths and beliefs.

As in other remote and underdeveloped regions of the world, the sun, moon and planets were used as the natural means of counting time and fixing the seasons. This resulted in a 13-month year, which required adjustments every few years in order to maintain a relationship between the months and years. It would appear that when it became apparent that the lunar months and seasons had gone out of phase, a meeting of elders would be convened. There would be long discussions which often led to renaming of months and sometimes the addition or deletion of months. It must be remembered that Africa is the second largest continent in the world. The equator runs across the middle of the continent, with two-thirds of the land

area to the north and one-third to the south. This
means that the skies give different indications of the
seasons in the north and the south. The appearance
of the Pleiades cluster signalled the time for crop
planting, becoming visible to the farmers of
Southern Africa in the morning and to the Masai
and Kikuyu in East Africa in the evening.
Throughout the continent the belief seems to have
been that the earth was the centre of the universe,
although there were differing theories about the sun.
There was a concept that the sky was a solid blue
rock resting on the earth and the stars shone
through holes in the rock. The sun was thought to
move across the solid sky each day and each night
return to the east. In some areas the sun and the
moon were thought to be little larger than the size
they appear to be in the sky. The moon was a much
more obviously changing object and, as in many
cultures, it became steeped in mythology. There
were various forms of moon worship throughout
Africa and the day of a new moon was often a day
set apart for rest or celebration. The brightest stars
in the sky and the most prominent of the
constellations were known and named as was Venus,
the most easily seen object in the night sky. It was
not always understood that the morning and
evening appearances of Venus were the same planet,
resulting in it often being given two names. Meteors
and comets were observed everywhere and were

generally thought to signify bad omens, such as the death of a chief, disease, famine or war. There were some exceptions to this belief, with the Masai of East Africa and some other groups believing that meteors were good omens which told of good rains coming. It is interesting to note that the ancient Africans did not leave buildings, memorials or artifacts bearing astronomical signs or representations.

It is thought that the Aborigines of Australia first settled in the vast island continent some 100,000 years ago. They were hunter-gatherers who had no real political organisation or unity, the only cohesive unit being the tribe. It has been said that, in spite of the fact that they had no instruments, their celestial observations were the most precise possible for a people using the naked eye. They had names for the sun and moon, the planets which were visible to them, comets and meteors, and also many of the stars. They were able to recognise various groups of stars and it is hardly surprising that they gave them names such as boomerang and kangaroo. As with other primitive cultures, the planets and stars entered the mythology of the people and many colourful and fanciful stories were told to explain the presence of certain stars and groups of stars.

The scattered islands of Polynesia were probably one of the last areas of the world to be reached by man. Although distributed over a huge area of the Pacific Ocean, the peoples of the islands were from a common stock and were culturally and linguistically related. The Maori of New Zealand had specialists whose job it was to study the skies, and this they did with the naked eye, like the Aborigines of Australia. They recognised and gave names to the sun and moon, the planets which they could see, the brighter stars and also the Milky Way. This level of knowledge was also reached by the people of the Polynesian islands. These islanders were very accomplished sailors who based their navigation on their knowledge of the stars. As with the Aborigines, the Maori and Polynesian people completely integrated their observations of the skies with their mythology and the accumulation of knowledge was passed from generation to generation by the telling of stories. Although the Aborigines, Maori and Polynesians did not contribute to the development of scientific astronomy, it is surely of great interest to know that, even in the most remote regions of the earth, human beings were able to come terms with their environment, learn to use the stars for their own needs and pass on this knowledge to future generations.

CHAPTER 2
The New Era

We return now to Europe where at almost the same time as Kepler was publishing his first two Laws of Planetary Motion, other great things were happening. In 1608, in Holland, the first telescopes were made and a whole new era dawned in the world of astronomy. The first systematic observer using a telescope was a great Italian mathematician, physicist and astronomer called Galileo Galilei. He was born in Pisa and legend has it that he demonstrated that the rate of fall of a body is dependent on its mass by dropping weights from the Leaning Tower of Pisa. When he learned of the invention of a simple telescope in Holland, he copied and then improved the design and was ready for his historic observations of the skies. The discoveries which he made were spectacular. He was the first to see the mountains and craters of the moon, the stars of the Milky Way, the satellites of Jupiter and, very significantly, the phases of Venus.

These phases could not take place if Ptolemy's theory with the earth at the centre was correct. He became convinced that the theory of Copernicus with the sun at the centre was correct and he wrote a witty and strong book, *Dialogue on Two World Systems*, in which he made Ptolemy's system look foolish. This book was published in 1632, only 16 years after the work of Copernicus had been condemned by the Roman Catholic Church as heresy and, inevitably, Galileo was accused of the same crime and brought to trial by the Inquisition. Under the threat of death, he was forced by the Inquisition to recant his views and was placed under house arrest for the rest of his life. It is said that he repeated aloud what the Inquisition ordered him to say, only then to mutter quietly, 'Still it moves'. It took until 1992 for a Vatican commission to clear him of heresy. Galileo was punished by the Inquisition in 1633, and yet the Ptolemaic theory was finally disproved in 1687, when Isaac Newton published his great work, *Philosophiae Naturalis Principia Mathematica*. In this famous and ground-breaking book, normally known simply as *Principia*, Newton formulated three laws of motion and laid down the laws of gravitation. It has been said that this book was the greatest mental effort ever made by one man. Newton explained the forces which hold the planets in their elliptical orbits and said that they also have a small but measurable effect on each other. He proved at last that the gravitational forces keep the earth and the planets orbiting the sun, that the earth is also rotating and that humans do not spin off the globe because of the same force of gravity. Newton, for the first time, made the proposition that the laws which govern the motion of objects on earth are the same laws which keep the sun and the planets in place and determine their motion and relationship with each other. He also, in 1666, separated the colours which constitute sunlight and this knowledge eventually led to the invention and use of spectroscopes, which are so important to the work of astronomers to-day.

The invention of the telescope brought dramatic changes to the study of astronomy and it opened up the way to more and more discoveries in the heavens. In Britain the principal use of the science was in navigation and in 1675 The Royal Greenwich Observatory was founded by Charles II in a building designed by Sir Christopher Wren. The meridian through Greenwich was adopted internationally as the prime meridian in 1884 and became the basis of Greenwich Mean Time. The first Astronomed Royal at Greenwich was John Flamsteed, who was set the task of compiling a catalogue of stars which could be used by sailors. He was followed by Edmond Halley who was the first to realise that comets do now appear at random but have periodic orbits. His name is perhaps best known because of Halley's

Comet, which he identified in 1705 and established that it appeared every 76 years. It last appeared in 1986 and will next be seen in 2062. Halley was a friend of Sir Isaac Newton and it was he who financed the publication of Newton's *Principia*. Another prominent name is that of Sir William Herschel, a German born British astronomer, who is often thought to be one of the greatest observers of all time. In 1781 he discovered the planet Uranus, the first such discovery since prehistoric times. He also discovered two new satellites of Saturn, binary stars and infrared rays from the sun. Sir William's son, Sir John Herschel, was also an astronomer and he continued his father's work of mapping binary stars and nebulae. In 1834 he set up an observatory in Cape Town and made the first detailed survey of the southern sky. Sir John was also very interested in photography and was involved at the beginning of the use of photography in astronomy. There were others active in the science at this time, including Charles Messier, the compiler of a famous catalogue of nebulae and star – clusters, and Johann Hieronymus Schroter, whose main work concerned the observation of the moon and planets.

CHAPTER 3
The Telescope

The arrival of the telescope on the scene was so important that it ushered in a whole new era in astronomy. No longer were the boundaries of the science fixed by the limits of human eyesight and conjecture. The surface of the moon could be studied in detail, showing the mountains and craters for the first time, and even the planets and the sun were brought into focus.

The first telescope was invented in Holland in 1608 by Hans Lippershey. The Dutch authorities attempted to keep the invention a secret but without success and the following year, in Italy, the initial design was improved and developed by Galileo. This was a refracting telescope in which the light falls on a converging long – focus objective lens. The image is then magnified by the short – focus eyepiece to produce the final image. The first reflecting telescope was produced by Isaac Newton in 1668. In this telescope the light from an object is collected by a concave, usually paraboloid, mirror of long focal length. The primary mirror reflects the light into a secondary optical system, which then reflects it into a short focus eyepiece. The lenses in the eyepiece create a magnified image which lends itself to being viewed by eye, photographed or analysed in other ways.

The two major types of optical telescopes are, therefore, refractor or reflector, although there are quite a few variations in design. As astronomers began to use this new tool which pushed back the boundaries of what could be seen and studied, their hunger for more and more detail and further and further exploration resulted in larger and larger telescopes. At Birr Castle in Ireland the Earl of Rosse built the largest telescope to date with a 72-inch mirror. It was a good telescope, although rather unwieldy, and the Earl had great success with it. Among other objects he found spiral nebulae which he reckoned were perhaps external galaxies, and were certainly very much further away than the first star which had been measured for distance. This was a dim star in the constellation of the Swan which had been measured by the German astronomer, Friedrich Bessel, at a distance of around 11 light years, or 60 million million miles. More and more refractor telescopes were built, the largest being the Yerkes Observatory in the United States with a lens of 40 inches, opened in 1897. There is a limit to the size which is possible for a lens for such a telescope and future development moved to reflector telescopes. The American George Ellery Hale set up a 60 inch reflector on top of Mount Wilson in California and in 1917 increased this to 100 inch. Using this telescope Edwin Hubble proved that the spiral nebulae which had been found by the Earl of Rosse were indeed galaxies far beyond our own, were travelling away from us at great speed, and that the whole universe is expanding. A large problem for astronomers is the atmosphere which surrounds the earth and, therefore, not only must telescopes be larger and larger – they must be higher and higher. The largest telescope in the world is now the one at a height of 14,000 feet on Mauna Kea in Hawaii. It has a segmented mirror with an amazing diameter of 387 inches. The ultimate telescope to date is, however, the Hubble Space Telescope which was launched in 1990 and has an orbit over 350 miles above the earth and has perfect conditions for observation at all times. It has a 94-inch mirror and was, quite rightly, named in honour of Edwin Hubble.

So much for the giant telescopes of the professionals. What type of instrument should the enthousiastic amateur be thinking of buying? It should be remembered that much enjoyable and useful work can be done with the naked eye. The field of vision which is possible with the naked eye provides the best basis for the initial survey of the skies and enables the new astronomer to plot the major objects and begin to understand their relationship with each other. When the planets and major constellations can be recognised with ease, it is then possible to select areas for more detailed observation with the aid of one of the tools of astronomy. Telescopes can be bought fairly cheaply these days but care should be taken in making the purchase as a great number of mediocre items are for sale in the mainstream high street shops. It will cost more to buy from a specialist outlet, but good advice is invaluable in selecting the telescope for you. The strength of magnification is of course important, but just as important is the aperture, which is the diameter of the objective lens in a refractor telescope or the primary mirror in a reflector. It is a good idea to buy the largest aperture you can afford as this will maximise the amount of light which the telescope will gather and will greatly aid the viewing of the dimmer objects in the night sky. Other points to be taken into consideration are the optical quality of the telescope, ensure that the eyepiece gives a good field of vision, there should be a detailed manual provided and, very important, check that the mount of the telescope is steady. The field of vision of a telescope is narrow and a shaky mount will make it very difficult to zero in on the area of the sky to be observed. The best and simplest mount for the first-time buyer is the altazimuth which can be adjusted both up and down and left and right. Before beginning to study the sky seriously, make yourself familiar with the finderscope attached to the side of the telescope. This gadget will make the targetting of subjects for observation much more easy and will save the frustration of weaving across the sky with the telescope in an attempt to locate something which you have spotted quite easily with the naked eye. The adjustments to the finderscope can be done in daylight and will not need to be repeated. Spend time understanding your telescope and read the manual carefully before beginning. Another good tip is to take about fifteen minutes to let your eyes become accustomed to the dark before starting observations. This will make all objects in the sky appear much brighter and enable you to locate and study those which are very faint. It is important to bear in mind that using a telescope can be tiring and it is helpful to take regular breaks to relax. Do not peer with one eye but rather, as with a camcorder, keep both eyes open, which is more relaxing and also steadies the vision.

Rather than buying a telescope which is too small and inadequate, a first class alternative is to invest in a pair of binoculars. A telescope gives an upside-down image but binoculars do not, and this gives them uses in other areas such as hill-walking and bird-watching, making them much more cost efficient. They are classified according to magnification and also the diameter of the main lens and, as they are hand-held, it is better to go for a 7-times magnification or thereabouts, as this level of magnification is easier to hold steady. As with a telescope, buy the largest aperture you can afford, the aperture being the diameter of the main lens. Check that the binoculars are comfortable to hold, that the focusing mechanism is smooth and steady and that there is anti-reflection coating on the lenses.

One of the greatest drawbacks today for the would-be astronomer is undoubtedly light pollution. Most of us are urban dwellers and we rarely have the opportunity to see the sky as it actually is. The author was recently in the north-west of Scotland and was able to see the sky, and particularly the Milky Way, with a clarity which he has not experienced for almost forty years! What can be done to minimise the effects of light pollution? Apart from travelling to the north-west of Scotland or the Kalahari Desert on a regular basis to make observations, it may be possible to take your telescope or binoculars to a nearby hill or try to go some way out of town. Make your observations in the opposite direction from the main light sources and you can buy telescope filters which will combat a good deal of the pollution from street lights. It will not always be possible to avoid light pollution, but there will still be much of the sky visible, providing interest and excitement for the amateur astronomer.

There is one warning which cannot be repeated too often to anyone who is looking at the sky. Do not look at the sun!! This is the only serious danger to the astronomer – but it is a serious one. Looking at the sun with the naked eye can be damaging but looking at the sun through a telescope or binoculars will be disastrous, the sun quickly burning the retina of the eye.

We have considered the invention and development of the telescope and its use by professional astronomers from the early days of Galileo to the Hubble Telescope in space. We have also come down to earth to consider the options available to the amateur astronomer in his choice of telescope or binoculars. We will now look at what is in space around us and learn more of our nearer and not so near neighbours.

CHAPTER 4
The Solar System

We have seen from our consideration of the beginnings of astronomy that the nature and history of the universe has always intrigued and excited man. He has been aware of the constantly changing positions and sense of movement of the sun, moon, planets and stars and has endeavoured to understand and give meaning to these strange celestial objects which obviously influence the earth, his home. It is not surprising that the sun, moon and planets entered human mythology and often took on the attributes of gods and goddesses. It is also unsurprising that man believed that the earth was flat and the centre of the universe. Many brilliant men gave lifetimes to the investigation of the mysteries of the universe, but it was not until the tremendous advances of the past three hundred years that real understanding took root. There are still conflicting theories regarding the origin of the universe and an immense amount of work is taking place on earth and also in space. New discoveries and theories regularly feature in news broadcasts and it is clear that the science of astronomy is growing and developing with each new move forward.

Many attempts have been made to establish how our solar system came into being, but the scientist historians have been hampered by the fact that the only evidence available to them has been evidence to be found here on earth. The knowledge of our system is now growing as space travel has enabled scientists examine samples of rock and other material brought back from the moon and various planets. Until now the strongest evidence has come from asteroids and meteoroids which have been found on earth. These relatively small fragments from space, as far as can be known, have reached the earth untouched since the origin of the universe and were the only pieces of evidence which astronomers could put their hands on. Modern techniques have been able to date these rocks and the results have been remarkably consistent. The ages of meteors, asteroids and pieces of moon rock vary between 4.4 and 4.6 billion years and, therefore, it has been deduced that the solar system came into being about four and a half billion years ago. There has been no shortage of theories regarding the origin of the solar system, but some form of agreement on the subject has emerged in the past 250 years or so. In 1755 the German philosopher Immanuel Kant proposed the theory known as the nebular hypothesis. This suggested that a nebula, a cloud of gas and dust, rotated slowly and, as it contracted, it flattened out into a spinning disk under the influence of gravity and from this disk was formed the sun and planets. A little later, in 1796, the French scientist Pierre-Simon Laplace proposed a similar theory which went further and suggested that the planets were pieces thrown off by the spinning disk and they then condensed, while the centre of the nebula formed the sun. Mathematical difficulties arose in Laplace's theory and it was eventually set aside by scientists, only for astronomers to return to his idea of a solar nebula in the middle of the 20th century. A modification of the original theory became known as the condensation theory and it assumed that the planets were condensed out of the cloud of matter surrounding the young sun and that they are all of similar ages. As the planets were being formed, those furthest from the sun were not subjected to the intense heat experienced by the planets closest to the sun, thus splitting the solar system into two distinct parts. The closest planets had the light gases such as hydrogen and helium driven off, making them solid and rocky, whereas the planets furthest from the sun and the heat retained their deep atmospheres of hydrogen and helium and are

not solid and rocky. These latest theories are where
the investigation of the birth of the solar system
stands at the moment. The investigation will
continue, but the evidence at present looks
convincing. So much for the birth of the solar
system, what about its death? This will probably not
occur for yet another four and a half billion years
and is, therefore, not a matter of immediate concern!
The life-span of the sun will most likely follow the
pattern of that of other stars of similar size. In time
the hydrogen at the core of the sun will be burned
up and, as this diminishes, the fuel in the outside
will also be used up, causing the outer shell to
expand and the sun to become brighter and brighter.
The sun will, in fact, become what is known as a red
giant, which is a star which is dying. The expansion
of the outer layers of the sun will affect the earth
dramatically as the mounting temperature evaporates
the water and atmosphere. It will eventually return
to its original arid condition, without life.
Meanwhile the sun will deteriorate into a burned-
out star, or white dwarf, and finally a black dwarf,
which is made up of mainly oxygen and carbon.

We now know that the probable life-span of the
sun, and therefore the solar system, is in the region
of nine billion years and we have touched on the
latest theories concerning the origin of the system.
Before looking at the components of the solar system
in more detail, let us take an overview of the

number of components involved and the immensity
of space in this system, which is only a corner of the
universe as a whole.

The sun, which is a star, is the centre of our solar
system and in orbit around the sun are nine planets
with at least 63 moons, more than 6,000 large
asteroids, a countless number of smaller asteroids
and comets, and billions of yet smaller meteoroids,
no larger than rocks. The distances involved are
staggering and very difficult to comprehend. The
earth is 93 million miles from the sun, Jupiter 483
million and Pluto 3,666 million. Astronomers often
try to explain these distances by reducing them by
scale and endeavouring to get the reader to visualise
something recognisable. For instance, if the sun was
a football, the earth would look like a pea some 165
feet away and Neptune, the farthest major planet,
would look like a grape at a distance of 7,400 feet,
or well over a mile. In order to be able to dicuss
these mind-blowing distances in almost a short-hand
fashion, two important astronomical units of
measurement are used. An Astronomical Unit, or
A.U., is equivalent to the average distance between
the earth and the sun, some 92,754,170 miles. The
most distant planet is therefore to be found almost
40 A.U. from the sun. The other measurement is
the Light-year, the distance travelled by light in one
earth year. Light travels at 186,000 miles per
second and in a year it covers around

5,880,000,000,000 miles, just under six million million miles. Such a number of zeros illustrates why some form of short-hand had to be devised. The star nearest to the sun is about 4.3 light-years away and the Pole Star 680 light-years, which means that we are actually seeing the Pole Star as it was 680 years ago. Perhaps it is no longer there, and if that is the case, we will not know for another 680 years! The solar system, although so vast, is itself a part of a much larger system known as the Galaxy. Around 100,000 million stars make up the Galaxy, the shape of which is often described as a double-convex lens. It appears to look like a disk which bulges at the centre and thins out at the ends. The centre of the Galaxy is in the bulge, known as the Galactic bulge, and our solar system is to be found in part of the thinner area, the Galactic disk, at a distance of between 25,000 and 30,000 light-years from the centre. When we are able to look along the plane of the Galaxy into the disk, we then see the beautiful band of glowing light which we know as the Milky Way. The stars in the Milky Way almost seem to merge but in fact they are widely spaced and the appearance of proximity is caused by the perspective from our position in the Galaxy. The Galactic disk contains a huge amount of dust and it is this dust which creates the dark band which runs the length of the Milky Way.

We have had a brief look at the composition of

the solar system and where it is situated in the Galaxy, of which it is a part. We will now examine the various components of the system and their relationships with one another.

CHAPTER 5
The Sun

The sun is a fairly average star in a galaxy containing somewhere in the region of 400 billion stars, but it is the master of the solar system and is by far the dominant component of the system. It is 100 times larger than the earth, with a diameter of 865,000 miles and its mass is more than 300,000 times more than that of the earth. In physics, mass is a physical quantity expressing the amount of matter in a body. It is a measure of a body's resistance to changes in velocity and also of the force experienced in a gravitational field. The sun is made up of incandescent gas which is approximately 73 per cent hydrogen, 25 per cent helium and the remainder containing small amounts of other elements such as nitrogen, oxygen, carbon, magnesium and iron. All of the elements found in the sun are also found in all other stars and bodies across the solar system and the universe. The solar energy is produced at the core of the sun by nuclear fusion reactions and is then moved to the surface, from where it is radiated into space as heat and light. The surface has a temperature of between 5,000 and 6,000 degrees centigrade and at the core the temperature is thought to reach an incredible 14,000,000 degrees centigrade at least. This hot central core has a diameter of around 250,000 miles. The fuel for the nuclear fusion reactions is the vast amount of hydrogen present in the sun. The nuclei of hydrogen atoms come together to create the nuclei of helium, which is the next lightest element, taking four hydrogen atoms to make one helium atom. This is a very efficient method of producing energy and accounts for the longevity of the sun. The energy generated by the sun would require around 600 million tons of hydrogen to be fused into helium every second, a fact which underlines the immensity of the store of hydrogen held in the sun.

The most visible part of the sun as seen from earth is not solid but is a layer of the sun's atmosphere, known as the photosphere, and this is the layer which produces the most light. Above the photosphere is the chromosphere, which normally cannot be seen against the brilliance of the photosphere. The chromosphere is largely made up of hydrogen, which gives it a pinkish colour which can only be seen from earth during a total solar eclipse as a pinkish aura around the solar disk. Higher than the chromosphere in the atmosphere of the sun is the corona, which is also only visible during a total solar eclipse, when it appears as a luminous crown around the dark disk of the sun. The highest temperatures in the atmosphere of the sun are to be found in the corona and these remarkable temperatures heat the gases to such an extent that they are able to escape the huge gravitational pull of the sun and a stream of electrons and protons are shot out into space. This is known as the solar wind. The earth is protected from this wind by the magnetic field surrounding it but some of the particles become trapped in certain areas of the atmosphere called the Van Allen Belts, which are named after the man who discovered them. Some of these particles rain down on the earth's polar regions and cause displays of colour and light as they hit the atmosphere. In the northern hemisphere these are known as the Aurora Borealis, or Northern Lights, and in the southern hemisphere as the Aurora Australis, or Southern Lights.

The atmosphere of the sun is subject to very violent storms and these produce results which can be seen on earth. Most of the activity on the sun is studied by telescope but some can be seen by the naked eye, recorded as far back as the time of Ancient China. Sunspots were reported regularly after the invention of the telescope at the beginning of the seventeenth century and Galileo's observations of them and their movements led him to conclude that the sun rotated. The sun does rotate, but not in the same way as the earth or other solid planets. The equatorial region rotates faster than the poles, known as differential rotation, and this is similar to the rotation to be found in what are called the jovian planets, Jupiter, Saturn, Uranus and Neptune. Because of this rotation, a sunspot will move slowly across the side of the sun facing the earth before disappearing and coming into view again in about two weeks time. It is now known that sunspots are caused by the sun's magnetic field, the lines of force breaking through the surface of the photosphere and bringing about cooling. The sunspots show as dark areas on the face of the sun and usually appear as pairs or groups, although they are only temporary phenomena. Small single spots may also appear, although they are normally of short duration, some showing for only a few hours. In 1843 a German astronomer called Heinrich Schwabe discovered that there was a fairly regular solar cycle and that the sunspot activity reaches a maximum on average every 11 years. Sunspot activity was very low during the 1980's and 1990's but maxima did occur in 1958, 1969, 1980 and 1991. Another factor in a cycle is that the first sunspots appear at the higher solar latitudes and later spots occur closer to the equator, without actually breaking out on the equator itself. The total sunspot cycle can, therefore, be said to be 22 years. At the maximum of the sunspot cycle, there are often violent explosions of gas on the surface of the sun called prominences and flares. Prominences tend to take place over a period of days or even weeks, whereas flares explode with tremendous power and release of energy in minutes or, at most, in hours. It is the flares which shoot out the electrified particles which create the polar lights and can also affect radio reception on earth. Sunspots can easily be tracked as they move across the face of the sun but flares and prominences are now usually studied by using the principle of the spectroscope. The sunspot activity takes place in the photosphere and above this lies the chromosphere, which can also be examined by the use of spectroscopic equipment. The corona, or outer atmosphere, however, can only be seen properly during a total solar eclipse.

Total eclipses of the sun have been seen and noted from ancient times and often had mystical connotations. Even in the times of ancient China eclipses were forecast with remarkable accuracy and today, not only is an eclipse an event eagerly watched by the general public, it is an opportunity for professional and amateur astronomers alike to see more clearly the upper atmosphere, and especially the corona, of the sun. There are three types of solar eclipse. A partial eclipse is when only a part of the sun is hidden and an annular eclipse is when the moon is furthest away from the earth, causing a ring of the sun still to be visable around the dark shape of the moon. The total eclipse comes when the moon passes between the earth and the sun and exactly lines up between the two. The diameter of the sun is 400 times greater than that of the moon but, amazingly, the two look the same size from earth because the sun is 400 times further away. This means that during a total eclipse the moon covers the complete disk of the sun, leaving only the sun's atmosphere to view. It is then possible to see the chromosphere with the red of the flaming prominences and also the wonderful corona, which appears as a luminous crown. The corona is very elusive and can only really be seen at the time of a total eclipse, which is why astronomers use the few minutes of an eclipse to the best possible advantage and will travel to any part of the world to be in position and ready for the great event. The Saros is the period of about 18 years between eclipses. Although not an exact period of time, the Saros is a very good indication of when the next eclipse will take place. It has been said that the Greek, Thales of Miletus, used the Saros to predict the eclipse which took place on 25th May 585 BC. The last total eclipse to be seen in the U.K. was on 11th August 1999, when it passed over Cornwall before crossing the Channel to France. There was great media excitement and Cornwall had a great influx of professional and amateur astronomers as well as the general public to witness this facinating and dramatic natural sight. On television we were able to watch the shadow of the moon sweep across Cornwall, northern France, western Europe and on eastwards.

So much is still unknown about the sun and study continues day-by-day, aided now by probes in space which enable astronomers to view parts of the sun which are difficult to see because of the angle which we have. This is particularly true of the polar regions of the sun. What we do know of the sun is awesome. Its size, its sheer power and the energy which it produces from its immense heat, without which the earth would not exist and could not survive, make it the most important body in the sky to us earthlings. However, we must not forget that, in the context of the whole universe, our sun is only an average star. It is close enough to us for us to be able to study it in detail and so learn more and more about the other stars which surround our solar system and stretch out to the furthest limits of the universe. As amateurs, however, we must not finish this chapter on the sun without stressing once more the danger involved in studying it in detail. Do not look directly at the sun through a telescope or binoculars as the light and heat will be focused onto your eye and you will be blinded, possibly permanently. Even using a dark filter is not safe and the only recommended method is to aim the telescope at the sun, without looking through it, and project the image on to a white card or screen. This is safe and works very well.

We have examined the sun, the originator and provider of the solar system, and we will now consider the other members of the solar system family, who are our near and not so near neighbours. We will start with our own satellite, the moon.

CHAPTER 6
The Moon

It is small wonder that the moon has facinated and intrigued man since the beginning of time. Apart from the sun, the moon is the largest and most obvious object in our sky and it has been said that while the sun rules the day, so the moon rules the night. Over the years the moon has been held in awe, it has been worshipped, it has regulated the timing of planting and harvesting, it has lit up dark nights, and it has been sung about romantically as often as June! Because of the relative proximity of the moon, it was possible to observe many aspects even before the invention of the telescope. The monthly cycle of the moon was well recognised, if not perfectly understood, as were the changes in its face and the human resemblance of that face. When the telescope was invented and Galileo had adapted it for astronomy, almost certainly the moon would be his first target for investigation and, without doubt, he was the first human being to look at the moon through a telescope. What Galileo literally discovered through the telescope was that most of the popular conceptions of what the moon was like were wrong. Instead of being completely smooth, it was, in fact, extremely rough and pock-marked. Like the earth, the moon had mountains and valleys, craters and what at first looked like seas. These discoveries raised the possibility that the moon was perhaps more like the earth than had ever been dreamed. If there was nothing very different about the moon then, perhaps, there was nothing different about the planets, or even the stars. Could it be possible that there was nothing extraordinary about the earth itself and that it was merely another body in space? Galileo also found the terminator, which is the name given to the boundary between the day and night hemispheres of the moon and which slowly crosses the moon during its orbit of the earth. He came to the conclusion that the heights of the mountains were similar to those on earth and by using geometry he attempted to calculate the altitudes.

The moon as viewed from earth is as large as the sun and appears to be a major player in the solar system. Nothing could be further from the truth. The moon is less than one third of the size of the earth with a diameter of 2,158 miles and it only has 1/80th of the earth's mass. This low mass has been instrumental in shaping the moon as the

gravitational pull is much less than that on earth,
resulting in the lack of ability to retain atmosphere.
Whether this has always been the case is not known,
but today the moon is airless, waterless and lifeless.
The origin of the moon is still very much a topic of
debate. Over the years there have been various
theories which have been discarded such as the
moon being a part of the earth which spun off to
form a new planet. Other theories include the
possibility that the earth and the moon came into
being at the same time as a two planet system, or
that the independent moon at some time came too
close to the earth and was captured by the earth's
stronger gravitational pull. All of these theories have
been dismissed for various physical and
mathematical reasons and there would now seem to
be increasing agreement with the idea that the birth
of the moon was caused by a collision. The theory
is that in the early evolution of the solar system the
earth collided with a huge celestial body, perhaps as
large as Mars. The earth would have been in the
process of forming and, therefore, still molten. The
impact would have caused a vast amount of debris
to spin into space and parts of the earth then
combined with the other body to form the moon.
The earth and the moon are around the same age, as
has been proved by the rocks brought back by the
Apollo astronauts and this would seem to support
the latest theory.

The moon moves round the earth and it is in synchronous orbit, meaning that the time taken by its orbit is exactly the same as the time taken by its own rotation. This means that the moon always shows the same face to the earth. The orbit of the moon is not circular and the distance from earth varies between 221,460 and 252,700 miles. The closest distance is known as the perigee and the furthest as the apogee. The moon takes 27.3 days to orbit the earth. Although the earth is the dominant body and its much greater gravitational pull keeps the moon in orbit around it, the moon also has gravitational effects on the earth. The greatest of these is the tidal effect on the oceans of the world. The gravitational force of the moon has been strong enough to cause the earth to be slightly misshapen. This is because according to Newton's law of gravity distance alters the force. The gravity of the moon affects the parts of the earth closest to it more powerfully than the parts which are at a greater distance and this has resulted in the earth being egg-shaped, rather than a perfect sphere. An easier target than the solid mass of the earth is the water on the globe. The gravity of the moon causes the oceans nearest to it to flow towards it, creating a high tide, and the same force is felt on the opposite side of the world in order to balance the globe. The movement of the waters at both sides of the earth are called tidal bulges and these bulges sweep round the globe as the moon moves round, giving every part of the world two high and two low tides in each 24 hours. The sun also has a gravitational pull on the earth, although much weaker than that of the moon because it is so much farther away. When the sun and the moon are pulling on the earth on the same plane at the time of a new or full moon, the tides are at their strongest and are known as spring tides. When the sun and the moon are pulling from different directions at the time of the half-moon, the tides are at their weakest and are known as neap tides. The use of the word spring in relation to tides has no connection at all to the season of spring.

The most obvious aspect of the suface of the moon are the craters which give it the pock-marked appearance. These craters have been caused by meteors crashing on the moon, which does not have the protection given to the earth by atmosphere, which burns up most meteors before they hit the ground. Like a well lived-in face, the moon shows very clearly every blow it has received. A collision with a meteor creates great heat which melts and blows apart the rock on the surface and leaves debris and dust. The debris displaced by a meteoroid impact is called an ejecta blanket. It is thought that the moon was subjected to very severe bombardment by meteors until around 4 billion years ago, when it appears to have eased off. Some time after this period, volcanic activity took place on the moon and the largest craters were filled with lava. There is still debate regarding the history of the moon and there is a body of opinion which reckons that the craters were produced by the volcanic activity and not the impact of meteors. The moon is as old as the earth but, unlike the earth, it would appear to be almost completely dead, with little perceptible seismic activity.

The phases of the moon are caused by the fact that it rotates round the earth and is reflecting different amounts of sunshine at different times. When the moon comes almost exactly between the earth and the sun, the light of the sun is striking the other side of the moon which we never see, and it is invisible to us. This is the time of the new moon and soon the slender crescent moon is to be seen in the night sky. As the moon continues its orbit round the earth, more and more can be seen and this period is known as the waxing crescent until it reaches the first quarter, when one half of the moon is reflecting sunlight. The orbit then moves on to the end of the next quarter which is full moon and we can then see the full sunlit face. The moon moves on to the last quarter when the other half is reflecting and finally into the last part of the orbit, known as the waning crescent. When the moon is more than half but less than fully illuminated, it is said to be gibbous. This takes place in the periods between the first quarter and the full moon and the

full moon and the last quarter. When the moon is
in its crescent stages the dark side can at times be
seen to be shining faintly. this is earthshine, which
is light which is being reflected onto the moon from
the earth. The moon is an ideal subject for the
amateur, with plenty of features of interest without
the danger associated with study of the sun. It is
safe to look at the moon through a telescope or
binoculars. The moon can be viewed on any clear
night but the most dramatic views are usually to be
found during one of its crescent phases when the
low sunlight throws the rugged landscape into relief.
A very interesting exercise is to look at the moon
each night from the new moon onwards through all
of its phases. The cloud cover will, no doubt,
interrupt at times but in good weather it should be
possible to view most nights. With a good map of
the moon by your side you will be able to make
your own discoveries, but early in the waxing
crescent look out for the Mare Crisium and the
prominent craters Burckhardt and Geminus. At this
time the Mare Tranquillitatis will also be coming
into view. The so-called seas on the moon have all
been given Latin names and the Mare Tranquillitatis
is the Sea of Tranquillity of Apollo 11 fame. At
about the eighth day after the new moon, when the
moon is at its first quarter, is a very good time for
observations. After this the moon enters the waxing
gibbous phase and the light becomes too bright to

allow detailed observation and this is true also of the waning gibbous phase. As the moon begins to wane, better viewing conditions return and the Apennine Mountains can be seen clearly. These are the features which were studied by Galileo in great detail and it is thrilling to follow him and gaze on the mountains associated with this immense figure in astronomy. It is a good idea to get a map of the moon, as you will then become familiar with the terrain and also the names of the main features. The Italian astronomer Riccioli drew a map of the moon in 1651 and then proceeded to name the main craters. He used the names of famous people, mostly astronomers, and this tradition is still carried on. As Riccioli's map was produced in the mid seventeenth century, some of the most illustrious names in astronomy missed out in the christening of the major craters. Names such as Newton, Hubble, Herschel father and son, and Halley are either missed out or have been allocated to rather insignificant craters. Galileo was very badly treated by Riccioli, who did not agree with the his ideas, and his name was given to a remote crater in the Oceanus Procellarum, the Ocean of Storms. Galileo was obviously out of favour with Riccioli, who was quoted as saying that he had 'flung Galileo into the Ocean of Storms'. For the amateur there is a whole small world out there to be explored, and when you have dealt with the moon, there are still the planets and stars!

CHAPTER 7
The Space Age

This is now the Space Age. Since the dawn of time, men have gazed at the planets and stars and wondered. Many have dreamed of escaping the bounds of earth and travelling to the stars, and yet it was not until forty years ago that the first man broke free of the atmosphere of the earth and made the first flight into space.

As long ago as the second century AD Lucian of Samosata, a Greek satirist, wrote a story about a flight to the moon, called *The True History*, because, in his own words, it was nothing but lies from beginning to end. This told how a ship was caught in a water-spout and thrown upwards so violently that after seven days and nights it landed on the moon. Various fanciful methods of flying to space were suggested, discussed and even attempted over the years, and then Jules Verne wrote his novel *From the Earth to the Moon* in which three people were fired from a gun towards the moon. Most of Verne's facts were wrong, but he was correct with regard to the velocity required to leave the earth, seven miles per second. The faster an object is propelled upwards, the higher it will rise before falling back to earth and the speed needed to defeat the gravitational pull of the earth is approximately 25,000 miles per hour, or seven miles per second.

The idea of spaceflight excited scientists for centuries and it is interesting that three men, quite independently and unknown to each other, laid the foundations for the spaceflights which began to be made in the middle of the twentieth century. The first of these three was a Russian schoolteacher called Konstantin Eduardovich Tsiolkovsky (1857–1935). He was a shy person who was almost totally deaf and seems to have been a bit of a loner. His real interest was in flight and he actually built a wind tunnel, in which he tested various aircraft designs. His enthousiasm for flight progressed to spaceflight and he was the author of the first serious published work on the subject. In his work on rocket propulsion he recognised that solid fuels would not work and he proposed a proper rocket motor, using two liquid fuels which would be forced into a combustion chamber to produce the gas required to fire the rocket. Tsiolkovsky also put forward the idea of a two-stage rocket, which was to be used in actual spaceflight some thirty years after his death. The second of the three scientists was an

American, Robert Hutchings Goddard (1882–1945). In 1926 he successfully launched a liquid fuel powered rocket which did not rise very high but reached a speed of 60 m.p.h. This proved that Goddard's theory regarding the use of liquid fuel to power a rocket was valid. The third scientist was Hermann Oberth (1894–1989). He was wounded while serving in the German army in the First World War and during his convalescence he studied physics and aeronautics. While still in the army, he conducted experiments on weightlessness and also invented a long-range rocket powered by liquid fuel. He could raise no interest in his invention and eventually he published his ideas in a book in 1923, followed by another in 1929. His second book won a prize which helped to finance the launch of his first rocket in 1931. During the Second World War, Oberth worked with Wernher von Braun in the development of rocket weapons. It took a war to focus the attention of the countries involved on rockets and a total of about 3,700 V-1's and V-2's were launched against London and southern England. At the end of the war the Soviet Union and the United States of America rushed to get their hands on the rocket technology and the scientists who had developed it. Wernher von Braun and many others found themselves in the United States, working on the American rocket programme at Cape Canaveral in Florida.

There was feverish activity on both sides of what was now being called the Iron Curtain, and in 1949 the team in America had produced a scientific rocket which reached a height of 250 miles, compared to the 60 miles achieved by the wartime V rockets. The research continued and by 1955 the authorities said that it would be possible for a satellite to be launched in 1957 or 1958. On 4th October 1957 a three-stage rocket was launched from the Soviet base in Central Asia. It reached a speed of 17,000 miles per hour and placed Sputnik 1 into orbit in the vacuum of outer space. The 23 inch, 184 pound sphere completed its first orbit in 96 minutes, bleeping as it went. The Americans were stunned. Their attitude seemed to be that 'their Germans are better than our Germans' but this was not so. Sputnik 1 was a purely Russian achievement which was overseen by a Russian scientist who, at that time, was unknown to the Russian public. His name was Sergei Pavlovich Korolev and he had been working on the possibility of space travel since the 1930's. Korolev is known to have met Konstantin Tsiolkovsky, the Russian father of space travel whom we have mentioned earlier. The second Soviet launching, Sputnik 2, took place on 3rd November 1957 and this time it had a passenger, Laika the dog. Laika was the first living creature to be sent into space, and she was there to help the Russians study the effects of prolonged weightlessness in orbital flight. It was obvious that the Russians were planning to put a man in space before the Americans had even launched a satellite. At last, on 31st January 1958 the first American satellite, Explorer 1, was launched into space. The race in space had begun. Once the breakthough had been made, all types of satellite were launched, scientific, communications and, of course, military. The frenzied activity continued in both camps and, although earth satellites were sent up in great numbers, the real prize was going to be putting a man on the moon. The Americans had set-backs with their rockets and still the Russians appeared to be ahead. On 13th September 1959 they crash-landed Lunik 2 on the moon and on 4th October of the same year Lunik 3 circled the moon and televised man's first view of the hidden far side. The Americans had nine more unsuccessful attempts before Ranger 7 landed on the moon on 31st July 1964. Meanwhile the Russians had been perfecting the recovery of satellites and space probes, using animals as passengers. On 12th April 1961 the world was electrified to receive the news that the first man had escaped the earth's atmosphere and had been launched into space. Yuri Gagarin completed a full orbit of the earth before landing safely near Saratov, some six miles from the planned landing area. A woman who witnessed the landing asked if he really had come from outer space, to which he replied, 'Just imagine it, I certainly have'.

The space race had now become the moon race. The Soviet leader Krushchev was quoted as saying 'Let the capitalist countries try to catch up'. The reply came from President Kennedy on 25th May 'I believe that this nation should commit itself to achieving the goal, before this decade is out, of landing a man on the moon and returning him safely to earth. No single project in this period will be more impressive to mankind, or more important or more for the long-range exploration of space; and none will be so difficult or expensive to accomplish'. The United States was now commited to putting a man on the moon before 1970, and there is no doubt that the Soviet Union was doing everything in its power to beat them to it. Both countries were training the men (and in the USSR, also the women) to go into space, and the build-up and the preliminary flights were taking place at remarkable speed. Manned flights became almost commonplace, satellites were put into orbit and men walked in space, protected from certain death by only their spacesuits. These spacemen were known in the US as astronauts, and in the USSR as cosmonauts. 1967 was a year of disaster for both programmes. In January three astronauts, Gus Grissom, Roger Chaffee and Edward White, died in a fire which broke out in an Apollo capsule on the launch pad at Cape Kennedy, as Cape Canaveral was now called. In April Vladimir Komarov, the Russian cosmonaut died when his Soyuz 1 spacecraft crashed. These accidents held up manned space flights for more than 18 months. To add to these tragedies, Yuri Gagarin was killed when his ordinary plane crashed on a training flight.

The American Apollo programme was to triumph in the race to the moon, using the Saturn rocket designed by von Braun. The first manned flight after the fire disaster was Apollo 7 which, apart from practicing essential manoeuvres for the journey to the moon, gave the world remarkable demonstrations of weightlessness on television. Anyone who saw it will remember the toothbrush floating in mid air. Apollo 8 was the first flight by man to the moon and it also was memorable. On 21st December 1968 the crew of Frank Borman, James Lovell and William Anders were blasted from Cape Kennedy by the massive 363 foot Saturn V rocket. It was only the third time that this rocket had ever flown and the first time it had carried human passengers. On the afternoon of Christmas Eve they reached their destination and travelled behind the moon, out of sight from the earth. They started up the motor of the service module to slow down and reach a lunar orbit only 69 miles above the surface of the moon and, for the first time, the astonauts saw the moon in close-up. The astronauts were obviously deeply moved by their experience and on Christmas Day they sent back to earth pictures of the lunar landscape as they read aloud from the Bible. The Apollo 9 mission was launched on 3rd March 1969 to test the lunar module, which would separate from the command module to make the landing on the moon. In the words of the mission commander, the purpose of the flight was 'to see if this whole gigantic mess would be able to fly to the moon and land there'. During the flight the two spacecraft were separated and flew as far as 100 miles apart before re-docking. There were no problems and the lunar module was now ready for lunar flight. Apollo 10 flew on 18th May 1969 with the task of taking the complete Apollo spaceship to the moon for the final test before landing. Some hitches developed as they approached the moon and there was some concern at mission control, but they carried out the important final tests, took many cine and still photographs and flew directly over the landing site selected for the actual landing, scheduled for July. The three astronauts had taken many of the risks for little of the glory and had prepared the way for the first man on the moon.

It is not the intention in this book to go into the incredible engineering feats involved in the construction of Apollo 11, but perhaps some dimensions will help to bring home the enormity of the project. The first stage of the Apollo 11's Saturn V rocket, built by Boeing in New Orleans, was 33 feet in diameter and 138 feet long. It weighed 288,750 pounds but fully fuelled with liquid oxygen and kerosene this increased to over 5 million pounds. It was delivered to the Kennedy Space Centre by sea-going barge. The second stage, built by North American Rockwell in California, also had a diameter of 33 feet and a length of almost 82 feet. It weighed 79,918 pounds which increased to just over 1 million pounds when fully fuelled with liquid oxygen and hydrogen. It was also delivered by barge. The third stage, built by McDonnell Douglas in California, was just under 22 feet in diameter and a little over 58 feet in length. It weighed 25,000 pounds but this increased to 260,523 pounds when fully fuelled with liquid oxygen and hydrogen. This stage was delivered by an aircraft called the Super Guppy. The Apollo spacecraft consisted of the launch escape tower, the boost protective cover, the command module, the service module and the lunar module within the lunar module adapter. The total height of the spacecraft came to 82 feet, making a total height of the craft and the rocket of 363 feet.

All was now ready for the landing on the moon of Apollo 11. The three astronauts selected were Neil Armstrong, spacecraft commander, Edwin 'Buzz' Aldrin, lunar module pilot, and Michael Collins, command module pilot. All of these men were very experienced pilots of conventional jet aircraft and had been making flights into orbit for three years. They were all military flyers and, in fact, Aldrin had flown 66 combat missions in Korea and Armstrong had flown 78, being shot down once but managing to parachute to safety in friendly territory. All three were 39 years old.

At 2.32pm BST on Wednesday 16th July 1969 the countdown was complete and the voice of launch control announced 'zero, all engines running. Lift-off, we have lift-off'. Apollo 11 blasted-off, watched by 3,500 newsmen and almost a million spectators, with a journey ahead of it of exactly 218,096 nautical miles. The previous Apollo flights had tested and rehearsed the complicated manoeuvres required with the command module and lunar module, and at 7.12pm BST on 20th July the two separated with the following words –

Collins: 'OK Eagle, one minute until T. You guys take car'.

Eagle: 'See you later'.

While behind the moon, the lunar module dropped to a lower orbit and was ready for the final descent to the surface of the moon. The world listened with bated breath to the radio communications between mission control in Houston and the two astronauts in Eagle as the lunar module gently approached the surface, and then – 'the Eagle has landed'. The time was 9.18pm BST on 20th July 1969. About two hours later, Armstrong and Aldrin ate man's first meal on the moon before putting on their moon suits and making the final preparations for the first walk on the moon. At last the hatch of the LM was opened and Armstrong slowly descended the ladder, pausing at the foot. At the same time the television camera on board began to show the first vague black and white pictures. At 3.56am BST 21st July 1969 man made his first step on the moon, with the famous statement from Armstrong 'That's one small step for man, one giant leap for mankind'. Once they were on the moon, the astronauts described everything they saw and experienced in detail. This was not just a great adventure but also a serious scientific expedition to learn more about the moon and man's reaction to the immense differences in environment. The thought must also have been there that it was possible that the men might not return to earth, and they should relay as much information as they could. It is interesting to read transcriptions of their reports to mission control as they experienced weightlessness, gathered rock samples and described the condition of the surface of the moon. The following are examples of these reports:

Armstrong: The surface is fine and powdery. I can pick it up loosely with my toe. It adheres in fine layers like powdered charcoal to the sole and sides of my boots. I only go in a small fraction of an inch, maybe one-eighth of an inch, but I can see the footprints of of my boots and the treads in the fine sandy particles. There seems to be no difficulty in moving around, as we suspected. It's even perhaps easier than the simulations at one-sixth G that we performed in the simulations on the ground. It's actually no trouble to walk around.

Aldrin: OK, you do have to be rather careful to keep track of where your centre of mass is. Sometimes it takes about two or three paces to make sure you've got your feet underneath you. About two or three or maybe four easy paces can bring you to a nearly smooth stop.

Armstrong: It has a stark beauty all of its own. It's like much of the high desert of the United States. It's different but it's very pretty out here. Be advised that a lot of the rock samples out here, the hard rock samples, have what appear to be vesicles in the surface. Also I am looking at one now that appears to have some sort of phenocryst.

Vesicles are bubbles in the rock, which on earth would be associated with gases bubbling out of a hot lava to form rocks like pumice stone. Phenocryst is a mineral found in volcanic rocks on earth.

At 6.53pm that afternoon of 21st July the two astronauts lifted off in the top half of the lunar module, leaving the descent stage of the LM at Tranquillity Base. They rose to meet Collins in the command module in moon orbit, and so began their long journey home. Their stay on the moon had lasted 21 hours and 36 minutes.

Apollo 12 landed on the moon in November of the same year, but Apollo 13, which was launched in April 1970, suffered an explosion and had to return to earth before reaching the moon. The world held its breath as the three brave and skillful astronauts defied death in flying back to safety. In 1971 and 1972 there were four more successful moon landings, the last being Apollo 17 in December 1972. Unbelievably, at this time there was a decline in public interest in space flights and exploration of the moon. The decline in interest combined with falling financial backing and great improvements in the success of unmanned missions brought manned moon landings to an end. There have been no manned landings since 1972, although there are continuing advances in space technology, which will be dealt with later in this book.

CHAPTER 8

The Inner Planets

We will now consider the planets in the solar system, and it is usual to divide these into the inner planets and the outer planets. The inner planets are those which are closest to the sun and are the nearest neighbours to the earth, which is part of this group.

Mercury is the planet closest to the sun. It orbits the sun between the earth and the sun and it is about 36 million miles from the sun and 50 million from the earth. Mercury is much smaller than the earth, with a diameter of 3,030 miles, compared with the earth's 7,926 miles. It takes 88 days to orbit the sun and has an axial rotation period of 58.6 days. Because of its proximity to the sun, Mercury is a very difficult body to observe. It is never seen against a dark background and it is really only visible when it is low in the sky. A good telescope is required and the best times for viewing are just before sunrise or just after sunset. Even if

Mercury can be sighted, it is almost impossible for a telescope on earth to reveal details of the surface features. It was not until the mid 1970's that Mariner, an unmanned space probe, sent back pictures of the suface showing craters but also valleys and mountains. There is a huge basin, which is over 800 miles in diameter, and has been given the name Caloris Basin. The reason for the name is that when Mercury is at its closest to the sun, the sun is directly overhead the basin, producing a temperature of around 430 degrees centigrade. The Mariner probe detected a weak magnetic field around Mercury, suggesting that the planet has a large core rich in iron. It also confirmed that the planet, like the moon, was without atmosphere, which leaves it vulnerable to meteoroids, X-rays and ultraviolet radiation. No further probes have so far been planned and there would appear to be no possibility of a manned expedition in the foreseeable future.

Venus is the other planet which orbits the sun between the earth and the sun. It is about 67 million miles from the sun and 26 million from the earth. Venus takes 224.7 days to orbit the sun and its axial rotation period is a remarkably slow 243 days. The diameter of Venus is very similar to that of the earth, 7,523 miles compared to the earth's 7,926. Until quite recently, Venus was known as the 'planet of mystery', due to the dense cloud cover and thick atmosphere which prevented any possibility of observing the surface of the planet. Because it is so much closer to the sun than the earth, and the cloud cover is so heavy, Venus absorbs much more of the sun's energy and heat, without being able to radiate much of it away. This results in a very hostile environment of incredible heat and a harsh world without water. All of the planets as viewed from earth spin in a counterclockwise direction except Venus, which spins clockwise, and no reason has been found for this. Perhaps the most likely explanation is that the planet sustained a violent collision at the time of the formation of the solar system which caused its slow retrograde spin. Although Venus is the earth's nearest neighbour except for the moon, the mysteries surrounding it remained without explanation until 1962, when an American probe called Mariner 2 came within 22,000 miles of the planet and relayed the first ever reliable scientific information. Many of the theories which had suggested that Venus might be a pleasant place with water had then to be discarded. The radar maps which have now been drawn indicate that the surface of Venus is made up of vast sweeping plains and highlands, which only cover about 8 per cent. There are two main highland areas, one in the north called Ishtar Terra which is about the size of Australia, and one in the south called Aphrodite Terra which is similar in size to

Africa. The atmosphere consists almost entirely of carbon dioxide and the clouds are made up of sulfuric acid droplets. Since 1962 there have been other probes which have orbited the planet and have studied the surface by radar and, in fact, the most recent of these was Magellan which was launched in 1989 and relayed information until 1994. In 1970 the Russians managed to put one of their probes onto the surface of Venus. It exerienced difficulty in negotiating the dense atmosphere, but made the landing and was able to send back useful information before it stopped transmitting after half an hour. The rocks reflect light from the clouds and appear orange-coloured. In the official report, the Russians graphically describe the light on the planet as being similar to the light in Moscow at noon on a cloudy winter day. It was decided that all the features on Venus would be given feminine names, but not before the mountains next to Ishtar Terra had been named after the great Scottish mathematician James Clerk Maxwell. This is the only feature with a masculine name and the Maxwell Mountains rise to a height of well over 30,000 feet. Venus is the second brightest object in the sky after the moon and is normally easy to locate. It is much closer to the sun than the earth and is to be found in the direction of the sun. Look for it in the western sky at or just before sunset, or in the eastern sky around sunrise. It is a bright shining light which does not twinkle like a star, but shines steadily. It goes through phases similar to the moon and appears to be larger during its crescent phases, when it is closest to earth and is also backlit by the sun. Because of the thick atmosphere, it is not possible to see anything of the features on the surface, but if observations are made before the sky is completely dark or completely light, it is more likely that atmospheric features will be visible. The next planet further from the sun is, of course, the earth, which is similar in size to Venus but has a totally different environment and supports such a range of life of all kinds. We will not consider our own planet in any detail in this book, but it is, perhaps, sobering to reflect on the significance of the extra distance we have between us and the sun. If the earth had ended up nearer the sun at the birth of the solar system, we would not be here, and we would not be concerned about studying the other objects in the universe! Man is unlikely to attempt a landing on Venus, although further unmanned probes will probably be sent to learn as much as possible about our relatively near neighbour. No matter how much we learn, however, it is unlikely that it will become a welcoming destination.

The third of the inner planets is Mars and it is very different from Mercury and Venus. To begin with, Mars is further from the sun than the earth at a distance of just over 141 million miles. It takes

687 days to orbit the sun and its axial rotation
period is 24.6 hours, very similar to that of earth.
Another similarity to earth is that Mars is inclined
on its axis at an angle of 25.2 degrees, compared
with the earth's inclination of 23.5 degrees. Both
Mercury and Venus have axes which are almost
perpendicular to their orbital planes. Mars is very
much smaller than the earth, with a diameter of
4,222 miles compared with our 7,926. The
atmosphere is very thin and consists of 95 per cent
carbon dioxide, with 3 per cent nitrogen and 2 per
cent argon. Although Mars is often referred to as the
red planet, it is, in fact, very cold and very dry, with
a mean surface temperature of minus 23 degrees
centigrade compared with plus 22 degrees on earth.
The atmospheric pressure on the planet is only
about 1/150th of the pressure on earth and that
compares with Venus where the pressure is about
100 times greater than that on earth.

There has long been a feeling on earth that there
were many similarities between Mars and our own
planet and many of the theories were completely
wrong. The red appearance suggested heat, whereas
Mars is essentially cold. There seemed to be the
possibility of life similar to our own, which is now
disproved, and some of the features appeared to
indicate the presence of water, which has also been
disproved. The speculation over the years included
the theory that some of the markings on the surface

of Mars were canals used for irrigation of crops. The speculation about life on Mars almost seemed to fuel a real wish to find some kind of life there and there have many books and films on the subject. Unfortunately these hopes and wishes were dashed when, in 1997, the Mars Pathfinder probe landed on the surface and sent back exellent pictures. This was the culmination of a programme of probes over the previous thirty years which had built up a better and better picture of what the conditions on the planet were really like. The Mars Pathfinder showed a landscape of red rocky soil and it was then proved that the colour came from iron ore, the Martian surface containing large amounts of iron oxide. The various spacecraft which have been sent to Mars have shown mountains, valleys, ridges, canyons and huge volcanoes. One of these volcanoes, Olympus Mons, is the largest known volcano in the solar system, with a diameter of 340 miles and a height of almost 17 miles, or around 90,000 feet, three times the height of Mount Everest. Another feature on earth which is made to appear insignificant is the Grand Canyon in Colorado when it is compared with the Valles Marineris on Mars which in places has a depth of over 4 miles, four times as deep as the Grand Canyon. It is about 2,500 miles long and up to 75 miles wide. Because the axis of Mars is inclined at almost the same angle as that of earth, like the earth it experiences different seasons.

During the Martian summer winds, which play a big part in the weather and environment on Mars, blow the red dust, creating vast dunes and exposing the dark surface underneath which is heavily cratered. There are features which appear to be dried river beds and these are considered to be evidence that at one time there was water on Mars. This is the kind of evidence which keeps the thought alive that there may have been life and that there is some kind of life now, albeit on a microbial level. Mars has two moons or satellites named Phobos and Deimos, names which come from the horses which pulled the chariot of the Roman god of war, Mars. The names mean Fear and Panic. These are nothing like the earth's moon in size, being only around 20 and 10 miles in diameter and are probably large asteroids. They are both of irregular shape and are very close to their parent planet. Phobos is only about 3,600 miles above Mars and it orbits the planet in only 7 hours 39 minutes. It is calculated that it is not in a stable orbit and could crash onto Mars, although not for another 40 million years! Deimos is in a stable orbit and is further away from the planet. The best time to observe Mars is when it is closest to the earth and on the opposite side of the earth from the sun, which means that the earth is between Mars and the sun. These times are called when Mars is in 'opposition' and occur for a few months every 26 months. The planet can be seen at other times and a small telescope will be able to show the polar caps and the main dark areas. More powerful telescopes should be able to observe the changes in the polar caps and to follow the clouds and dust-storms. Plans are even now being worked on for the first manned flight to Mars and the target date is 2020. The technology is there for man's next foray into space.

CHAPTER 9
The Outer Planets

We will now travel away from the sun, past Mars, past the main belt of asteroids, to find the giants of the solar system. The outer planets are also known as the jovian planets, jovian meaning 'like Jupiter'. Jupiter, the ruler of Mount Olympus in Greek mythology as Zeus, and also the principal Roman god,controlled the weather and used the thunderbolt as his weapon. His name became Jove in Middle English, hence the adjective 'jovian'. After the sun, the planet Jupiter is the real giant of the solar system and the other planets in the outer reaches of the system have similarities to it and are often referred to as the jovian planets. In the same way, the inner planets are often called the terrestrial planets because of similarities which they have to the earth.

Until the invention of the telescope, Jupiter was thought to be merely a bright star, but in 1610 Galileo (who else?) looked with his new telescope and saw for the first time the banded surface of Jupiter and also its orbiting satellites. The jovian planets are totally different in make-up to the earth and the terrestrial planets and, although they are much larger and more massive than the earth, they are much less dense. They have a dense core but much of their size and mass is made up of surrounding gases. Jupiter is around 483 million miles from the sun and it takes almost 12 years to go round it. The diameter is 89,424 miles, compared with the earth's 7,926 and the sun's 865,000. It has an axial rotation period of 9 hours 50 minutes. From earth we can only observe the top clouds of Jupiter, but it is known that the thick atmosphere is made up of over 80 per cent hydrogen and the remainder is mainly helium. The atmosphere is reckoned to be over 600 miles deep and there are other gases in the mix, such as ammonium hydrosulphide, ammonia and hydrogen sulphide, all of them together producing a fairly obnoxious aroma! When the solar system was developing, the outer planets furthest from the sun were cooler and had greater mass, which enabled them to retain the hydrogen and helium in their atmospheres by gravitational pull. When observing Jupiter, the most obvious features are the dark belts across the planet and also the bright zones. These can be seen with a small telescope, as can the best known, and till recently, the most mysterious feature on Jupiter. This is the Great Red Spot, which was first observed over 300 years ago. It is an immense feature which can be found in a fairly central position just above the equator of Jupiter, in an area known as the Red Spot Hollow. It can appear at times to be grey, but it is often a very definite red. Over the years there have been many theories put forward to explain the Great Red Spot, such as a glowing volcano or some kind of solid object which at times would be visible and at other times would be submerged in the deep, thick atmosphere. It was not until recent times, when space probes investigated the planet, that it was found to be a jovian storm which whirls in an anti-clockwise direction for periods of 11 days at a time.

Our knowledge of Jupiter has really only grown since the advent of the space age. There were two space programmes involved in the investigation of the planet, Pioneer and Voyager. In 1972 and 1973 two Pioneers were launched, and after journeys of almost two years they both passed Jupiter and were able to relay useful information. The Voyager programme was, however, more successful in sending back new facts. Voyager 1 and Voyager 2 were launched in August and September 1977 and they reached Jupiter in March and July 1979. Voyager 2 passed the planet at a distance of 440,000 miles but Voyager 1 approached even closer, coming to a minimum distance of 217,000 miles. The Pioneers had established that there are strong areas of radiation around Jupiter which could be damaging for space probes and fatal for any manned flight venturing too close. The two Voyagers were therefore instructed to pass quickly over the equatorial areas, where the largest concentration of radiation is to be found. The clouds covering Jupiter were able to be examined in detail for the first time and the constant storms were shown to be a main feature of conditions on the planet, accompanied by thunder and lightning. The Voyager probes also discovered that there are rings round Jupiter, although they are not at all obvious. The next step forward in the exploration of Jupiter was the Galileo probe which was launched in October 1989. A much longer route was taken on this occasion and Galileo did not reach Jupiter until December 1995. The probe consisted of a craft to orbit the planet and also an entry craft. The two sections were separated before reaching the planet and the entry probe succeeded in sending back signals for 57 minutes as it plunged through the clouds to its destruction. The other section went into orbit around the planet and provided information which was completely new on Jupiter itself and also on its satellites. There is a total of 15 satellites, four of them major and the other eleven are small. The four major moons are known as the Galileans, as they had been observed by Galileo in 1610, using his first telescope. These four moons range in size from Europa, which is slightly smaller than our moon, to Ganymede, which is larger than Mercury and much larger than our moon. The other two moons are Io and Callisto. These moons can be seen by using almost any telescope or a good pair of binoculars and, in fact, the largest, Ganymede, can at times be seen with the naked eye. A sighting of Ganymede was recorded in China as far back as 364 BC. The space probes have shown Ganymede and Callisto to be cratered and dead like our own moon, Europa to be strangely smooth on the surface, and Io to be, by far, the most dramatic. Io is the closest orbiting moon to Jupiter at a distance of about 262,000 miles and has very active volcanoes and a surface which is red. It is reckoned that Io is so different from the other three because of the pull of the immense gravitational field of Jupiter which produces a tidal effect, thus creating the volcanic activity which spews large amounts of black lava at temperatures of around 350 degrees Centigrade. In common with the other major moons, Io has practically no atmosphere and it travels in the depths of the radiation produced by Jupiter, making it a very inhospitable world indeed. The other satellites circling Jupiter are very small indeed, with diameters less than 200 miles.

The next of the outer planets nearest to the sun is Saturn. Although it is the next closest, the distances involved are mind-blowing and almost impossible to comprehend. We have seen that Jupiter is about

483 million miles from the sun and Saturn is a further 403 million – a total of 886 million miles from the sun. The distances, of course, get much greater as we reach out to the other planets and beyond and we can understand how far this science has come from the days when men believed that the earth was the centre of the universe. Saturn has an orbital period of 29.5 years, compared to Jupiter's 11.9 years and, also like Jupiter, it is a quick spinner with an axial rotation period of 10 hours 14 minutes, compared to Jupiter's 9 hours 50 minutes. In make-up Saturn is in many ways similar to Jupiter, being mainly composed of hydrogen and helium gases, but its total density is lower. The core of Saturn is much larger than the size of the earth, with a temperature possibly as high as 15,000 degrees Centigrade. This tremendous heat contrasts with an extremely cold temperature in the top clouds of the planet of minus 180 degrees Centigrade. Saturn is considered to be perhaps the most beautiful object in the sky and the main reason is the spectacular rings which encircle the planet. These can be seen through a small telescope or binoculars and, although they appear to be solid, they are, in fact, composed of huge numbers of small pieces of ice spinning round the globe. At the right time the rings make a wonderful sight, and even the divisions between the two main rings can be seen. These rings, A and B, are separated by

what is known as Cassini's Division, after the Italian astronomer who discovered it in 1675. There are other rings round Saturn which can only be seen properly with a good telescope. As with Jupiter, space probes have added a great deal to the knowledge of Saturn. In 1974 Pioneer 11 sent back good photographs as it passed but, more importantly, Voyager 1 and Voyager 2, fresh from their surveys of Jupiter in 1979, reached Saturn in November 1980 and August 1981. Voyager 1 passed within 78,000 miles while Voyager 2 approached to 63,000 miles, such proximity being possible as the problem of radiation was not as acute as that experienced around Jupiter. Very good pictures of Saturn were sent back to earth, especially of the rings, and for the first time it was obvious that the system of rings was much more complex than had been previously thought. We now know that there are hundreds of lesser rings separated by narrow divisions and that there are also rings outside the main system. Many of these new discoveries are extremely faint and practically impossible to view with telescopes on earth. The successes in space exploration are pushing back the boundaries of knowledge of worlds millions of miles from earth. Saturn has a great many satellites, although only 18 have been definitely identified. Many are very small indeed, while seven are of medium size and 'tidally locked' with Saturn.

'Tidally locked' means that, because their orbits are synchronous, they always show the same side to their parent planet, as our moon does to the earth. The really significant satellite is Titan, which appears to be unlike any other world in the solar system. Titan is large, with a diameter of 3,201 miles and it orbits some 760,000 from Saturn. It is unique among planetary satellites as it has a substantial atmosphere which is thicker than that on earth. This atmosphere consists of 90 per cent nitrogen, almost 10 per cent argon and traces of methane and other gases. For comparison, the earth's atmosphere is made up of 78 per cent nitrogen, 21 per cent oxygen and 1 per cent argon. The nature of the surface of Titan is still unknown and the Voyager probes were not able to help with this. The temperature of the surface is very low, which has enabled Titan to retain its thick atmosphere, and the existance of this atmosphere has led to some speculation that there may be some form of life there. There is likely to be great excitement in 2004, when a space probe is due to make a controlled landing on Titan, hopefully answering many outstanding questions. This probe is to be in honour of the discoverer of Titan in 1655, Christiaan Huygens.

Moving on outwards from the sun, the third giant of the outer planets is Uranus. It is at a distance of 1,783 million miles from the sun and

orbits the sun every 84 years. It has an axial
rotation period of 17 hours 14 minutes. The
atmosphere of Uranus is made up of about 84 per
cent hydrogen, 14 per cent helium and some
methane and ammonia and it has a diameter of
31,770 miles. Uranus was only discovered in 1781
by William Herschel and it has not been possible to
gather too much information on the planet. Even
the pictures received from the Voyager missions
showed little detail of Uranus. The rotational axis is
inclined at an extreme 98 degrees, with the result
that there are periods when one of the poles is
facing the sun. Uranus has fifteen satellites, ten of
them being discovered by the Voyager 2 probe. The
largest of them is Titania and even it is less than
1,000 miles in diameter. The only satellite or moon
which is really interesting is Miranda, which has a
most unusual surface with ridges, plains, craters and
oval-shaped faults.

Yet another 1,010 million miles from the sun, at
a total distance of 2,793 million miles is Neptune,
which was only discovered as recently as 1846. The
astronomers who found this planet were Johann
Galle and Heinrich D'Arrest, who had been
systematically looking for it in a position suggested
by Urbain Le Verrier, a French mathematician. Le
Verrier had shown that Uranus was being pulled out
of position by the gravitational force of a more
distant and, at that time, unknown planet. Neptune

orbits the sun over a period of just under 165 years and has an axial rotation period of 16 hours 7 minutes. The atmosphere on Neptune is very similar to that on Uranus, made up principally of hydrogen and helium and it has a diameter of 31,410 miles, again very similar to Uranus. There are, however, certain significant differences between the planets. Although few prominent features can be seen on Uranus, even in the pictures sent back by Voyager 2, Neptune displays an interesting area known as the Great Dark Spot, which is reckoned to be a vast storm region. This was seen by Voyager 2 but seemed to have disappeared when the Hubble Space Telescope focused on Neptune some five years later. Neptune reflects a very clear blue colour, which is caused by the clouds of methane which are gathered high above the planet. Neptune has 6 minor satellites and one major moon called Triton. Triton has a retrograde orbit, which means that it spins in the opposite direction from all the other moons of the outer planets. Because of this and the fact that it does not orbit in the equatorial plane of Neptune, many astronomers think that Triton has at some time been involved in a collision or some other violent event. It is also possible that it was not part of the original system of moons orbiting Neptune, but was captured at a later time by the gravitational pull of the planet. One of the minor moons is perhaps notable for its misshapen

appearance. This is Proteus which is only roughly
like a sphere and is covered in craters.

We now come to the ninth planet and the real
maverick of the solar system, which was only
discovered as recently as 1930. Neptune was
discovered in 1846 because the orbit of Uranus was
being disturbed by the pull of an unknown body
and mathematical calculations by Le Verrier
pinpointed the area where the new planet Neptune
would be found. The astronomer Percival Lowell
was convinced that yet another planet must be
responsible for the continued irregularities in the
orbits of both Uranus and Neptune and produced
calculations to find it. Although Lowell was not
successful in completing the search, an assistant
astronomer at the observatory which he had
established in Flagstaff, Arizona discovered the
elusive planet in 1930. This was Clyde Tombaugh,
and he actually found the planet by examining
photographic images. The new planet was named
Pluto and, amazingly, its position was no more than
six degrees from where Lowell had said it would be
found. It is, however, now thought that Lowell's
accuracy came about more by luck than good
judgement. After study it became apparent that
Pluto was too small and did not have enough mass
to affect such massive planets as Uranus and
Neptune. Pluto only has a diameter of 1,444 miles
and it is some 3,666 million miles from the sun,

with an orbital period of almost 248 years and an axial rotation period of 6 days 9 hours. The orbit is so eccentric that Pluto comes closer to the sun than Neptune every 248 years or so. This was indeed the situation between 1989 and 1999. In 1978 the further discovery was made that Pluto has a moon which is more than half the size of itself. The moon has been named Charon and it has a diameter of 806 miles, orbiting about 12,214 miles from Pluto. We have said that Pluto is the maverick of the solar system and it is quite different from all the other planets, both terrestrial (inner) and jovian (outer). It is possible that Pluto and its moon Charon are either debris left over from the original formation of the solar system, or that they came into being as a result of a collision among the moons of Neptune. Although Pluto can be seen by using an 8 inch telescope, it only shows as a distant star. Astronomers have not given up hope of discovering Planet X, which is the planet thought by Lowell to be affecting the orbits of Uranus and Neptune, and the subject of his calculations. However, no evidence has so far come to light, and Neptune and Pluto at this time form the outermost frontier of the solar system.

Before finishing our brief look at the planets of the solar system, it may be of interest to consider a strange formula known as Bode's Law or the Titius-Bode Law. In the late 18th century a German astronomer called Titius came across what appeared to be a relationship between the various distances of the planets from the sun. His theory was taken up by Johann Elert Bode of the Konigsberg Observatory and, as a result, the formula is normally referred to as Bode's. What Titius noticed was that if the distance between the earth and the sun is expressed as 10 units and 4 is added to a certain range of numbers, then the distances of each of the planets from the sun can be given fairly accurately. The range of numbers is 0, 3, 6, 12, 24, 48, 96 and 192. Examples of the perceived accuracy are as follows:

Venus by Bode's Law is 7, actual distance 7.2
Saturn by Bode's Law is 100, actual distance 95.4

At the time of the drawing up of this formula Saturn was the most remote of all the planets and astronomers began to use the law to try to discover planets even further away. Unfortunately, the law did not work for Neptune and it became apparent that the relationship between the numbers was pure coincidence.

CHAPTER 10
Solar System Debris

The sun is the centre of the solar system and all the planets, including the earth, orbit round the sun. The planets and their moons account for a good many celestial objects, but there is much more whirling around in the system. The formation of the sun and the planets resulted in a vast number of smaller objects which are among the oldest objects in the solar system. For a long time astronomers paid little attention to the smaller denizens of the system, concentrating on the more obvious targets for study, the sun, the moon and the planets. In recent times, however, there has been more work done on asteroids and comets and it has been realised that the moons and asteroids are fruitful objects for study. These are indeed the debris of the solar system but, because they lack the atmospheres and geological activity of the planets, they are likely to reveal more information on the origin and development of the system. More than 6,000 asteroids have been observed and catalogued, and those are the ones with regular orbits. The shape of the orbits do vary, with many being more oblong. These are called eccentric orbits and some can be very pronounced indeed. For instance, 5365 Damocles swings from the inner solar system out into the outermost reaches of the system, crossing the orbits of Mars, Jupiter, Saturn and Uranus on the way. Most asteroids are concentrated in the area between the orbits of Mars and Jupiter, known as the asteroid belt. A remarkable fact is that all known asteroids orbit in a similar way to the earth and the planets, with the exception of only one, which has a retrograde orbit. Asteroids are mostly made up of iron, carbon and rocky material and it is thought that this is the combination which came together at the formation of the solar system. These asteroids have been able to avoid the changes which have taken place on the planets and can be regarded as the fossils of the solar system. Launched in 1989, the Galileo probe had to pass through the asteroid belt as it headed for Jupiter. It was able to image two asteroids from close range, both of them looking rather like potatoes. One of them, 243 Ida, is 35 miles long and has its own moon, which is a rock rather less than one mile in diameter. This moon, which has been named Dactyl, is the smallest satellite in the solar system. Although most asteroids can be relied on to remain in the asteroid belt, those with eccentric orbits mentioned above do cross the orbits of planets and also the earth. Those which cross the path of the earth are known as Apollo asteroids and scientists are concerned that a collision is a possibility. A theory which many accept states that the extinction of the dinosaurs came about as a result of an asteroid impact which took place about 65 million years ago. Some scientists think that an asteroid struck somewhere in the area which is now Mexico, causing a vast cloud of dust to travel over the world, destroying many of the plants on which the dinosaurs depended for life. Although such an impact is not predicted within the foreseeable future, scientists believe that a system should be set up to give advance warning of such a possibility and that a defence could be mounted by nuclear missiles, which would either blow up an incoming asteroid or divert it to ensure that it misses the earth. It is thought that there could be collisions with asteroids of diameters of more than half a mile and that these would cause devastation similar to that caused by several hydrogen bombs. It was predicted in 1998 that the earth would have a close encounter with an asteroid around 2030. Doubt was raised about this possibility, but it is again being talked about as a risk which will have to be faced. Objects smaller than asteroids do hit the earth from time to time and in 1908 what may have been a small comet struck a remote region of Siberia, making an explosion which was felt over 600 miles away and destroying a huge area of forest.

The comet has come to have a reputation for heralding unusual events and being spectacular. It is then a bit of a let-down to find that comets are made up mainly of ice and dust or rubble. They are often described as 'dirty snowballs' and the name 'comet' comes from the Greek, which means 'hair', a passable description for the long tail of a comet. A comet is fairly small, never more than a mile or two in diameter, and this nucleus comprises the total comet for the largest part of its orbit. It is as it approaches nearer the sun that the rising temperature evaporates the ice in the nucleus and creates the head, or coma, and also the tail. The coma is a ball of gas and dust, and can measure perhaps 60,000 miles in width around the nucleus, while the long tail is made up of the same material. A bright comet, its long tail sweeping across the sky, is one of the great events in the heavens, but as the orbit then takes the comet away from the sun again, the coma and the tail disappear and the comet returns to its inert state. The evaporation which produces the coma and tail of the comet means that its mass is reduced each time it passes through perihelion, the closest it comes to the sun. Very often, comets which have been observed on a regular basis disappear and this is caused by the fact that their mass has been reduced and reduced, until they disintegrate.

There are two areas of the solar system where comets have gathered, the Kuiper Belt and the Oort Cloud, both having been named after the Dutch astronomers who discovered them. Comets which reach the inner system at intervals of less than 200 years are thought to come from the Kuiper Belt and are known as short-period comets. The comets which arrive at intervals of hundreds and even thousands of years are known as long-period comets and are thought to come from the Oort Cloud. Comets in these areas will only come within our ken when their orbits are disturbed by some outside gravitational influence and they are kicked into an eccentric orbit, taking them out of the belt or cloud. This new orbit lasts indefinitely, which means that the comet will continue to visit this part of the solar system on a regular basis until it disintegrates due to the reduction of its mass caused by each successive close approach to the sun. In 1705 the British astronomer, Edmond Halley, published a book in which he suggested that comets which had appeared in 1531, 1607 and 1682 were, in fact, the same comet. Previously it had been believed that comets passed the sun once only and never returned. Halley went on to predict correctly that this comet would be seen again in 1758, which it was. Halley had died before this happened and the comet was duly named in his honour. Halley's Comet appears every

76 years and further calculations have shown that it was first observed in 240 BC and was seen also in that important year in British history, 1066. The last visit of Halley's Comet was in 1986, when it was met by a real welcoming committee! Five space-craft were launched to meet the comet, two from Russia, two from Japan and one from Europe. The European probe was built in Britain and was named Giotto. It penetrated the coma of the comet and sent back images of the nucleus, which is mainly made up of ice and dust. The total width was approximately nine miles and it is covered with a porous black crust. An intriguing statistic is that each time this comet reaches its perihelion, or closest distance from the sun, about 300 million tons of its material is lost, thus ensuring that its life will come to an end at some stage. The comet travels at a speed of about 80,000 miles per hour and the tail is millions of miles long. Halley's Comet is the best known and most studied comet in the sky and it is due to visit us again in 2062. There have, of course, been many other outstanding comets over the years, including one in 1744 which had at least six tails, the Great Comets of 1811 and 1843 which could be seen in broad daylight, and Donati's Comet in 1858 which displayed a straight gas tail and a curved dust tail. This last comet has been said to be the most beautiful ever seen. The two brightest comets in the 20th century were to be seen in 1996

and 1997. The Hale-Bopp became very bright in the early part of 1997 and could be seen with the naked eye for several weeks. It is stunning to think that it will not be seen again for around another 4,000 years. Hunting for comets is a popular field of activity for many amateur astronomers and can be done with powerful binoculars or a wide-field telescope. As the object of the exercise is to find something unusual in the night sky, it is necessary to have a good knowledge of the star constellations as a background to the search. This is a time-consuming operation which is best done in a rural area, well away from the lights of town. It is said that the best time to search is just before dawn around the time of the new moon. What is required is great patience and suitable precautions against the cold.

There is a very strong connection between comets and meteors, which produce the most exciting sights in the heavens. Most people have seen a shooting star or a falling star, which are meteors and not stars at all. With a little knowledge, meteors can be seen on a fairly regular basis and the sightings do not require a telescope or binoculars. The term 'meteor' is used for the streak of light in the sky and 'meteoroid' for the rocky object which causes the light. Meteors are quite different from comets. Comets are very far away from earth, with the result that they appear to move slowly and can often be visible for months. They are also in orbit and will return again and again to be seen from the earth. Meteors are debris from comets or asteroids which are only visible as they enter the earth's atmosphere and are, therefore, very close to earth. The streak of light comes from the friction of the incoming meteoroid as it passes through the atmosphere and the heat which this friction generates. The largest percentage of meteoroids are completely burned up as they pass through the atmosphere, but a small percentage do survive the journey and reach the earth. The objects which land on earth are generally known as 'meteorites', although the majority of them do not come from meteors but from the asteroid belt and are, in fact, small asteroids. Great numbers of these meteorites have been found and some have even been seen during their passage through the atmosphere. On Christmas Eve 1965 a meteorite was observed in the sky above England, eventually breaking up and causing pieces to land around the village of Barwell in Leicestershire. There were no injuries on the ground and to date there have been no reports of death or injury caused by landing meteorites anywhere. Over thousands of years, some very large meteorites have struck the earth. Luckily it does not happen too often or the human population would be under great threat. Most of the action appears to have taken place in prehistoric times and there are very significant craters and actual meteorites in various parts of the world. The largest known meteorite is still to be seen where it hit the earth near Grootfontein in South Africa. It is estimated that the weight of this huge object is over 60 tons. The explorer Robert Peary found a large meteorite in Greenland and it is now to be seen in the Hayden Planetarium in New York.

Showers of meteors occur on a very regular basis and are caused when the earth's orbit takes it through the tail of a comet. The earth ploughs its way through the debris, or clusters of micrometeoroids, and the meteor shower is the result of the fragments burning up in the upper atmosphere. Because of the regular appearance of certain comets, many meteor showers can be seen at the same time every year and can be observed with the naked eye. These showers are so much part of the annual pattern of the skies that, not only are their periods of maximum activity accurately noted, they have also been given names. These names are related to the constellation of stars from which they appear to come. The meteors which appear in August come from the constellation Perseus and are called the Perseids, the ones which come from the constellation Leo are called Leonid. The radiant is the position in the sky from which the meteors in a shower seem to diverge and the name is derived from the constellation in which the radiant can be seen. Details of the most frequent showers are given in the appendices at the end of the book. The showers are classified according to the number of meteors which can be seen with the naked eye in ideal conditions and this is called the Zenithal Hourly Rate, or ZHR. The estimated hourly counts of the regular meteor showers range from about 10 to over 500, but the earth meets the densest part of of the tail of Leonid every 33 years, when the count rises from the average of 10 to an amazing 1,000 meteors per hour! These streaks of light in the sky are produced by very small meteoroids, some hardly bigger than a pea. It is apparent that comets and meteors are relatively easy for amateurs to observe and study, and can provide real interest and excitement. It is also surely important that so much can be learned about the solar system and beyond from what is the debris of the system. The comets and meteoroids bring information from afar, direct to our own planet.

CHAPTER 11
To The Stars

In the introduction to this book we referred to the importance of stars to the souls of ordinary people and poets. They have been a source of inspiration and consolation, aspiration and pride. We have film stars, football stars, pop stars and stars of stage, screen and radio. Those in love have stars in their eyes, stars are used on national flags, and we know with what pride children bring home their school work adorned with gold stars. Why have stars always been a manifestation of excellence? It is surely because they are both beautiful and inaccessible. Let us look at how inaccessible.

The distances which we have considered within the solar system are immense and, in all truth, almost impossible to comprehend. The distance between the earth and the sun is 93 million miles, hard to imagine. The distance between the sun and the furthermost planet, Pluto, is 3,666 million miles, impossible to imagine. We are now about to look at the stars, an altogether different league in relation to distance. Before learning of the shorthand terms used to express distance in relation to stars, let us use old-fashioned miles to give some kind of idea of the vast space which we are discussing. The star 61 Cygni in the constellation of the Swan is relatively close to earth, and yet it is reckoned to be about 64 million million miles away. How can distances of this magnitude possibly be measured? The answer was worked out by a German astronomer, F.W. Bessel, in 1838 when he decided to use the tried and trusted geometrical method of triangulation, as used by surveyors. This method means taking a baseline and then joining the ends of the baseline to the object which is being measured. By calculating the angles, the distance from the baseline to the object can be arrived at by mathematics. When trying to measure the distances to stars, however, the lines from the ends of the baseline are so long that it becomes impossible to measure the difference in the angles, as they will both be very close to 90 degrees. Bessel solved this problem by lengthening the baseline in a very clever way. He plotted the position of a star at an interval of six months, which meant that the earth had moved from one extent of its orbit to the other, giving a baseline now of 186,000,000 with which to work. The observations are made against more distant stars in the background and it is possible to trace the apparent shift of the star being measured relative to these stars. This apparent movement is known as stellar parallax, and by measuring this, the angle relative to the baseline is known and the star's distance is calculated by triangulation. This stellar parallax is the basis of one of the shorthand terms used to express distance in relation to stars. A parsec is the measurement of distance between the sun and a particular star when the parallax of that star is one arc second, or 1 degree. A simple way of understanding the idea of parallax is to hold your finger up at the end of your extended arm in front of a more distant object. If you look through one eye and then the other, the object behind will appear to move. The distance of apparent movement is the parallax. A light-year is the distance travelled by light in one year, somewhere under 6 million million miles and a parsec is 3.26 light-years, approaching 20 million million miles. Every star is further from the sun than even this mind-numbing distance! Even the astronomical unit (A.U.) which is the distance of the earth from the sun, some 93 million miles, is much too cumbersome to be used in expressing distances outside the solar system. Even the use of parsecs has its limitations as the maximum distance at which the parallax can be realistically measured is around 100 parsecs, or approximately 333 light-years, or very approximately 2,000 million million miles. This is as far as the best telescopes in the world can help with the measurements. It is again staggering to know that the majority of the stars in our own galaxy are farther away than this distance. If the majority of stars are too far away to be measured even by the parallax method, how can astronomers judge the positions and distances of stars? By comparing what is known as apparent magnitude with what is known as absolute magnitude. The idea of apparent magnitude or brightness goes back to Hipparchus in the second century BC. He classified stars into a scale according to their brightness to the human eye, ranging from 1 for the brightest visible to 6 for the faintest star in the sky. The same system is in use today, albeit in a refined form. The difference between the degrees of magnitude has now been set at around 2.2 to 2.5 and when this difference is taken into account it means that a magnitude 1 star is 100 times brighter than a magnitude 6 star. The system has also been adapted to accommodate stars which are brighter than magnitude 1 by creating minus degrees. This means that Sirius, the brightest star in the sky, has a magnitude of -1.5, while the faintest stars have a magnitude of +6. It is reckoned that the naked eye, on a clear night, can see stars to a magnitude of +6 and that binoculars can reach to a magnitude of +9. The most sophisticated of the largest telescopes in the world can now reach a magnitude of near +30.

While Sirius is the brightest star in the northern sky, the brightest in the southern sky is Canopus. The apparent magnitudes of these two stars are not very far apart, with Sirius at -1.5 and Canopus at -0.7, but these figures do not tell the whole story. Sirius is about 8.5 light years distant from the sun, whereas Canopus is roughly 1,000 light years away. The real luminosity and power of these two stars cannot be judged by only considering their apparent magnitudes as, in fact, Sirius is perhaps 25 times as powerful as the sun, compared to Canopus with 200,000 times the power of the sun. It must always be borne in mind that the stars which we gaze at in the sky are not the neighbours they seem to be. Although they make patterns with which we become familiar, each star is at a different distance from us and they would only be seen in these patterns from our perspective. The patterns will eventually alter because every object in the universe is moving, but this is something which need not trouble the readers of this book as the present shape of things will not change for thousands of years. Even the sun moves, and it is reckoned that the entire solar system is moving round the centre of the galaxy, taking about 250 million years to complete the orbit. We have established that apparent magnitudes are not reliable markers for estimating the vast distances from stars to the earth, and so the theory of absolute magnitude is used. In order to grade the luminosity of stars in a way which is not distorted by the differences in distance, a standard measure of 32.6 light-years, or 10 parsecs is used. The absolute magnitude is therefore the apparent magnitude of a star as seen from a standard distance of 32.6 light-years. This produces different values of magnitude. For example, the magnitude of Sirius moves from -1.5 to +1.4 and that of Canopus from -0.7 to -8.5. If we were, therefore, able to view the stars according to their absolute magnitudes, Canopus would appear as the brightest object in the sky after the sun and the moon. The apparent magnitude is the feature of stars which can be recognised from earth and a list has been made of the 21 brightest in the sky, these stars being said to be of the 'first magnitude'.

CHAPTER 12
Anatomy of The Stars

In the previous chapter we discussed the vast distances involved in reaching out to the stars and the methods used to determine these distances and the relative luminosities of the stars. We will now consider the make-up of these far distant bodies.

As our sun is a star, we have a good idea of the essential composition of the other stars, and this has been covered in some detail in an earlier chapter. Using the astronomer's modern trusty friend, the spectroscope, greater detail of distant stars can be deduced, aided by information which is already available on the sun. Normal types of stars show spectra broadly similar to that of the sun, although there are quite important differences in detail. Observation of the stars shows that there are differences in colour and these changes are related to the temperature of the star. In the way of astronomers, the stars have been classified according to their spectral types and the system has become a little complicated. The original intention was to place the hottest stars under type A, diminishing in temperature through types B, C, and so down the alphabet. There were complications leading to changes, but the essential classification is still in place. Each type has also been sub-divided to allow for even finer gradations. The hottest stars have surface temperatures up to 80,000 degrees C and the coolest around 2,500 degrees C. The table of stellar spectra is included in the appendices at the end of the book. Stars are also arranged according to size and this can be determined from the luminosity and surface temperature. A dwarf is a star with a radius similar to or smaller than the sun, a giant has a radius 10 to 100 times that of the sun, and a supergiant has a radius more than 100 times that of the sun. There are stars which have a radius as large as 1,000 times that of the sun. At the beginning of the 20th century, two astronomers produced diagrams which placed the stars in relation to their luminosity and spectral type. Strangely, the astronomers, Ejnar Hertzsprung from Denmark and Henry Norris Russell from the United States, worked quite independently on this. Their work was brought together in the Hertzsprung-Russell Diagram, known normally as the HR Diagram. The diagram shows quite clearly that the vast majority of stars fall into a well-defined area of the graph, highlighting the fact that the hottest stars are the most luminous and the cooler stars the least luminous. It was initially thought that the diagram described the life of a star from being large and cool at birth to being smaller and hot before dying. This theory has been disproved, but the diagram is still a very useful piece of work. The area of the diagram showing the majority of the stars is known as the Main Sequence. The 10 per cent of stars falling outside the Main Sequence are thought to be in the process of dying.

The life of a star is finite. As with the sun, a star is born with a certain amount of fuel and when that fuel is exhausted its life comes to an end. It is now known that the critical attribute of a star is its mass, which produces the power, and this is also affected by speed at which the power is used. Hydrogen is the main source of fuel, and the nuclear reactions which take place within the mass of the star produce the heat and cause the star to shine. The more luminous a star is, the more fuel it requires and the more quickly it uses up its fuel. This fact means that stars which have large masses are more luminous, use up their fuel more quickly and, as a result, have a shorter life. Low mass stars are not nearly as bright, require less fuel and use it more sparingly, resulting in a longer life. Although we talk of long and short lives for stars, we are, of course, considering huge extents of time. Stars which have perhaps 10 times the mass of our sun, but are thousands of times more luminous, will have a lifespan of around 20 million years, compared with the sun's 10 billion years. Proxima Centauri, a close neighbour of the sun, has only one tenth of the sun's mass and produces only a fraction of the sun's luminosity, but it can be expected to exist 100 times longer than the sun. It can, therefore, be summed up that stars with high mass have short lives and stars with low mass have long lives.

A star begins its life as a coming together of interstellar material and the type of life which it can expect is determined by its mass. Somewhere in the region of one tenth of the mass of the sun would seem to be the determining point for whether the fledgeling body becomes a full-blown star or is fated to be no more than a faint dwarf with a limited supply of fuel. When the mass becomes more than one tenth of the mass of the sun, then a real new star is born. It will still be millions of years before the core temperature reaches the right level to set off the nuclear reactions which are essential for the life and the luminosity of the star. The required temperature is in the region of 10 million degrees centigrade. The main fusion reaction process is core hydrogen burning, in which the hydrogen is fused into helium, thus creating great energy in huge quantities. In time the star will reach maturity and take its place with the vast majority of stars in the Main Sequence, where it will remain for as long as it retains the balance between the pressure of the fusion pushing outwards and the gravitational force pulling inwards. An average G-type star, like our sun, will continue to operate fully for a period as long as 10 billion years before beginning to deplete and finally run out of the hydrogen necessary for the nuclear reaction. When the supply of hydrogen is exhausted, the make-up of the star alters and the core accumulates more and more ash from the fusion which is slowing down. At the same time the balance between the fusion and gravity has been disturbed and gravity is winning, with the result that the star shrinks and hydrogen outside the core begins to burn. For a relatively short time the star is in a very strange state, with the core continuing to shrink and the outer layers burning and swelling. The size grows to many times the original and the star becomes much more luminous. A star the size of our sun is known as a red giant at this late stage of its life and more massive stars are known as supergiants. The red colour is caused by the cooling of the surface temperature. When the sun reaches this stage in about five billion years, it is thought that the luminosity will increase by about 2,000 times and the outer layers will swell out to roughly 0.7 A.U. This will bring the sun to within 30 million miles of the earth, causing the atmosphere and the oceans to boil away, leaving nothing more than a rocky planet.

The ageing star now moves on to the next stage in which it is known as a white dwarf. The outer layers have been totally blown away, leaving only the core, which has become amazingly dense. The tremendous forces which have been at play during the contraction of the core and the destruction of the outer layers have crushed the atoms and produced the unusual density, which supports what is left of the star from further collapse. The theory is that, in time, all light and heat will desert the white dwarf and it will therefore cease to shine and move on to become a black dwarf. The twist to this situation is that we are not able to see a black dwarf and cannot verify such a stage in the death of a star. It is by no means certain that any star has actually had the time in the life of the universe to become a black dwarf. The crushing of the components during the contraction of the core means that the protons and electrons combine and the result is a star which is totally made up of neutrons, a very strange star indeed. In size it is probably no more than a few miles in diameter, but the density is remarkable. There is a powerful magnetic field and the neutron star spins, possibly as fast as many times a second. In 1967 a radio astronomer at Cambridge University, Jocelyn Bell Burnell, came across a radio source which was making a ticking noise quickly and regularly. After some time it was deduced that the noise was being produced by a rapidly spinning neutron star which was emitting beams of radiation from its magnetic poles, and these beams were transmitted to the earth each time they crossed our path. These are pulsars, and hundreds of them have been located since 1967, mostly by the radio signals. When pulsars were first detected, the regular pulses raised the possibility of intelligent messages from outer space, but this theory was fairly quickly dispensed with! The pulsars are usually produced as the result of a supernova, which is an immense explosion caused either by the destruction of a white dwarf which is part of a binary system, known as Type I, or the total collapse of a star of great mass, known as Type II. It is a Type II supernova which results in pulsars. Only four supernovae have ever been recorded in our galaxy, the last being in 1604. The final phenomenon related to the death of stars which we will consider is the black hole. This is a concept which is difficult to imagine and takes place when a star is so massive that, as it collapses and the core is shrinking, the gravity of the immense mass takes over and the collapse continues for ever. A black hole allows nothing to escape and even light itself disappears. The boundary of the black hole is called the event horizon and it is not known what happens inside this boundary. The life of a star is a finite thing, even if it does take thousands or millions of years to run its course, and we still have much to learn.

CHAPTER 13
Special Stars and Groupings

A cursory look at the sky gives an impression of millions of individual stars shining and twinkling in competition with one another, but this may not be entirely the case. A large proportion of stars seem to travel the skies in company with other stars and, in fact, individual stars like our sun may be in the minority.

There are many double stars, or binaries, which are two star systems in which the partners orbit one another. Some of these systems can be seen with the naked eye, and others appear to be single stars until they are examined by telescope. Another category of binaries are so far away that they cannot be seen as two distinct objects, even with the most powerful telescopes. These systems are known as spectroscopic binaries and they are observed by what is called the Doppler effect. This effect means that a source of radiation which is coming towards an observer emits shorter waves than a source which is going away from the observer. As double stars orbit one another, the difference in wavelength of one approaching and one receding informs the astronomer that there are two stars in the system. If one star happens to be covering the other, the spectral lines of the visible star will indicate the presence of the other. The best known of all the double stars is Mizar, also called Zeta Ursae Majoris, which can be found in the handle of the Plough. This is a second magnitude star and it is accompanied by a very much fainter star called Alcor. Although these two are partners in a system and are light-years apart, they are following the same path through space and they would appear to have come from a common origin. It has been mentioned before that stars which appear to be close neighbours, and members of the same group, are often many millions of miles apart and only seem to be close because they happen to be on the same line of sight, as seen from the earth. The same is true of double or binary stars. An example of this is Theta Tauri, in the Hyades cluster, near Aldebaran. The two partners have 338 seconds of arc between them and can, therefore, be seen with the naked eye. They have almost the same magnitude, but observation with a telescope shows that one of the stars is white, while the other is orange. The orange star is almost twice the distance of the white from earth. If these two stars were viewed from another angle, they could well then appear to be at opposite ends of the heavens. Some double stars can be very colourful and beautiful when viewed through binoculars or a telescope. Examples of this are Albireo in the cross of Cygnus which is yellow and blue, and Antares in the Scorpion which is red and green. Another feature of some binaries is that the partners can be quite different from one another. The very bright star, Sirius, is part of a binary system and it is 10,000 times more luminous than its partner, which is also very small and very, very dense. Another binary system with a story is Castor, which is the fainter member of the Twins. Both partners of the Castor binary are themselves spectroscopic binaries, as is a much fainter member of the group. This means that there are, in fact, six stars making up Castor. There are other examples of multiple star groups.

An important binary is Algol, which is in Perseus. Algol is a star of the second magnitude which begins to fade every two and a half days. It fades for five hours, remains at the reduced magnitude for 20 minutes, and then takes another five hours to build up again to its original magnitude. This strange variation in magnitude was not noted until the seventeenth century and it was eventually explained by a young astronomer, John Goodricke. Algol was not a variable star as had been thought, but was an eclipsing binary. One partner is much brighter than the other and, when the fainter of the two comes between the other star and the earth, the luminosity diminishes until the eclipse passes. John Goodricke was a remarkable young man who died when only 22 years old, having had to deal with the major handicaps of being both deaf and dumb. He it was who discovered a class of genuinely variable stars known as the Cepheids, called after the first of the class to be discovered, Delta Cephei, which can be found in the north of the sky. This class of variable star has longer periods between the changes in brightness and the luminosity is greater than the other variables. Stars in this class tend to have periods of more than 50 days. The Cepheids are pulsating stars and the period of pulsation is dictated by the time that a vibration takes to move from the surface of the star to the centre and return to the surface. The larger and more luminous stars take longer to vibrate and thus have longer periods between changes of luminosity than the smaller stars. Study of the Cepheid variables by an astronomer at the Harvard College Observatory, Henrietta Swan Leavitt, produced a real advance in the early years of the 20th century. She had been observing variable stars in the Magellan Clouds, which are neighbours of our own galaxy. The stars in this area are all approximately the same distance from earth and Leavitt observed that the more luminous variables had longer periods than the fainter stars, and that it was possible to measure the brightness and the period of a star and from these calculations the distance to the star could be worked out.

There are many types of variables to be seen in the sky and the observation of these is now an important area for the amateur astronomer to explore. The professional astronomers cannot cover all of the possible sightings and welcome the help of the keen amateur. Other unusual occurrences are novae. Despite their name, these are not new stars but are double systems which team up an ordinary star with a white dwarf, which is extremely dense. It is thought that the white dwarf attracts material from the other star, which creates instability in the system, leading to a sudden explosion. The novae activity can last for weeks, months or even years and can be very spectacular. In 1918 one of these outbursts flared to a remarkable magnitude of zero. Other notable novae were seen in 1934 and 1975. Some stars have been known to have nova type explosions on more than one occasion, one such taking place with the same star in 1866 and 1946. This star, known as T. Coronae, was normally only of the tenth magnitude, but in 1946 it reached the magnitude of 2. On very few occasions, the accumulation of the material around the white dwarf can result in the complete destruction of the binary system in an explosion of catastrophic proportions.

When this happens it is known as a type I supernova. No living astronomer has had the experience of witnessing a type I supernova, as the last one to be seen was in 1604.

Another rich field for amateur astronomers are star clusters, many of which are well within the range of small telescopes. There are two main types of cluster, open clusters and globular clusters. One of the best-known of the open clusters has been recognised since prehistoric times and has had many legends attached to it. This is the Pleiades, or the Seven Sisters, which is a strikingly beautiful star-cluster of seven stars which are visible to the naked eye, in addition to other components. Observers with good eyesight can see more than seven stars and some have claimed to see nineteen. Many more can be seen with the aid of binoculars, and there could be a total of several hundreds. It is interesting to note that binoculars are of more use in studying clusters than telescopes because of their broader field. Quite a few clusters had been recognised before the invention of the telescope, and then in 1781 a French astronomer called Charles Messier produced a comprehensive catalogue of over one hundred objects which were commonly mistaken for comets. He was a hunter of comets who had often been frustrated by thinking he was observing a comet when, in fact, what he was seeing was faint objects well beyond the solar system. Eventually a

Danish astronomer called Dreyer compiled a more complete catalogue which became known as the New General Catalogue, or NGC. Both catalogues are used regularly and objects are often known by two catalogue numbers, for instance Praesepe is M44 as well as NGC2632.

Globular clusters are quite different from open clusters. Whereas open clusters are usually comprised of no more than a few hundred stars and have no regular shape, globular clusters can contain more than a million stars and are symmetrical. Globular clusters, unlike the open variety, are old and do not contain the material which could create new stars. They also include Cepheid variable stars, and in 1917 an astronomer called Harlow Shapley discovered how distant the Cepheids are by observing the periods. This gave him the distances to the clusters, and because of their positions in the southern hemisphere of the sky, he deduced that the solar system is placed towards one edge of the galaxy, and not in the middle. It is now recognised that the solar system is almost 30,000 light-years from the centre of the galaxy. The centre of a globular cluster is densely populated with stars, but the stars at the edges of the cluster are more widely apart. In such an arrangement, there will be little darkness, especially in the centre where the stars are only a few light-days apart. The brightest of the globular clusters are to be seen in the southern hemisphere, but the best which can be seen from Britain is M13, which can be found between Eta and Zeta Herculis. M13 was discovered in 1714 by Edmond Halley, and although it is difficult to see with the naked eye, it can easily be seen by using binoculars. The Hubble Space Telescope has provided more information on the globular clusters and it has identified unexpected newer looking stars in the centre, called blue stragglers. it would appear that these have been formed when two stars have come together to form a binary system. The main star feeds on its less dense partner, heats up, and then becomes a blue giant for the second time.

Nebulae are vast clouds of dust and gas in space. There are a good number of these clouds and many of them are immense, for example, M42, the Sword of Orion is about 1,500 light-years away and its diameter is reckoned to be 30 light-years in diameter. That is a diameter of about 175 million million miles! We are again dealing with distances which we cannot comprehend! There are bright nebulae and dark nebulae, the only difference being that the bright nebulae happen to have hot stars within the system or very close, and these either are reflected or they may ionise the hydrogen and emit light.

CHAPTER 14
Galaxies and Deep Space

During the course of this book we have moved out from the sun to the planets and the stars, and now we will push even further outwards to other galaxies and deep into space. Our own galaxy was touched on as we considered the stars, but another look will be useful before launching out to the other galaxies and beyond.

The popular name for our galaxy is the Milky Way, and a most beautiful sight it is. A bright carpet of stars which arcs across the sky, which looks rather fuzzy when seen with the naked eye but a stunning array of thousands of individual stars when seen through a telescope. Unfortunately the light pollution of the modern urban world in which most of us live means that very many people have never had the opportunity to enjoy this magnificent spectacle. In spite of the vast distances involved, the Milky Way is our home territory and a little time spent in learning about this, our own galaxy, will

surely help when considering the many other
galaxies in the universe.

We have already noted that the sun, the centre of
our solar system, is almost 30,000 light-years, or
175 million million miles, from the galactic centre.
The solar system is orbiting the galaxy and takes
about 225 million years to make one orbit, the
period of the orbit being known as the cosmic year.
Seen from the viewpoint of the solar system, the
galaxy appears to be shaped like a disk, which
bulges at the middle and thins out at the edges, with
the solar system located in the thin area, or galactic
disk. It is not possible for observers on earth to see
through the galaxy because of the dust in the way
and, in fact, it is possible to see much further in the
opposite direction to the galaxy, out into space. The
galactic bulge consists of old stars and stretches for a
few thousand light-years from the centre of the
galaxy. The disk which surrounds the bulge is made
up of a mixture of old and young stars, and also
contains gas and dust. It is about 1,000 light-years
thick and extends about 50,000 light-years from the
centre of the galaxy. The centre of the galaxy lies
beyond the star-clouds in Sagittarius and, although
it cannot be seen from earth, radio waves and infra-
red radiation can penetrate the dust and, in time, it
should be possible to learn more. Through radio
astronomy, it is known that the galaxy is in the form
of a spiral with arms, and that the sun is situated

close to the edge of one of the arms. These days it is possible for astronomers to learn more of the size and structure of our own galaxy by studying other galaxies in the universe, but this has only been able to be done in recent times.

It was in 1923 that Edwin Hubble, using the Mount Wilson 100 inch telescope, made an important discovery. Although Messier had catalogued many interesting objects in the sky, including two different types of nebulae, it was unclear whether these were within our system or outside of it. One of the types of nebulae appeared to be clouds of gas, the other seemed to consist of stars. The Earl of Rosse in Ireland, who has been mentioned earlier in this book, in 1845 found that many of the starry type nebulae were spiral in formation. They were not all of the same shape, but they were obviously fairly similar and were certainly very different from the other type of nebulae. More was now known about these interesting objects but where were they? It was Hubble who answered that question. Conscious of the work done by Henrietta Leavitt earlier in the 20th century on the law covering the periods and luminosity of Cepheid stars, he studied these spirals and specifically looked for Cepheids. He did, indeed, find them and measured their periods, from which he was able to estimate their distances. He estimated that the Andromeda Spiral was about 750,000 light-years away from the sun, thus proving that it was certainly not part of our galaxy but was a completely separate system, another galaxy. Some years later further discoveries led to these figures being updated, but Hubble's discovery still stands as one of the most significant ever made in astronomy. Once more astronomers had to revise their thinking. They now knew that the galaxy was just another galaxy, in the same way that it had been learned over the years that the earth was just another planet and the sun was just another star.

Having established that these very distant nebulae were individual galaxies, Hubble set about splitting them into different catagories with various subdivisions. Without going into too much detail at this stage, there are three broad catagories which are defined by shape: spiral, elliptical and irregular. Hubble graded the spiral galaxies into small, medium and large, with an extra division for what are called barred-spiral galaxies. These are galaxies which have a bar of stars running through their centre. The bar lies on the plane of the disk and the arms of the spiral start at the end of the bar. The elliptical galaxies look like fairly round balls of stars, bright at the centre and becoming more faint and wispy at the edges. Hubble arranged these into eight classes according to shape, ranging from almost circular to elliptical. All other galaxies fall into the miscellaneous catagory of irregular. These galaxies have no discernable structure and are very untidy looking. They are rich in interstellar material and are, therefore, good breeding grounds for new stars. Those astronomers who are observing in the northern hemisphere do not have the opportunity to view the most spectacular of the irregular galaxies, only visible in the southern hemisphere. These are the Small and Large Magellanic Clouds, named after the explorer Ferdinand Magellan, whose crew observed them during their round-the-world expedition.

When discussing galaxies it is important to bear in mind that the most amazing distances are involved, and it is therefore interesting that all of the research which has taken place indicates that these distant are very similar to our own galaxy. The usual suspects are there, planetaries, variable stars, open and globular clusters, nebulae and novae. On occasion supernovae will flare up, as in our own galaxy, and they can be seen across these vast distances, such is their power. In recent years amateur astronomers have had a good deal of success in looking for these supernovae, and this success has been appreciated by the professionals. The closest galaxies to our own are the Clouds of Magellan, at a distance of less than 200,000 light-years, and the Andromeda Spiral at 2.2 million light-years. As the Magellanic Clouds are a mere 1,200,000 million million miles away, they can be

studied in greater detail than the rest. Our galaxy is
part of what is known as the Local Group,
containing galaxies of varying sizes, and beyond this
group there are other clusters of galaxies, many of
them much larger than our own. One such group is
the Virgo cluster, about 50 million light-years away,
and this is possibly the centre of a huge number of
systems known as the Local Supercluster. The
measurements and calculations made on all these
bodies strongly suggests that all of these distant
galaxies are moving away from us at great speed.
This indicates that the whole universe is expanding
in all directions.

Many people have heard of quasars without really
knowing what they are. In the early 1960's radio
astronomy was becoming a strong tool in astronomy
and the radio signals which were being picked up
from all parts of the universe were being
documented and catalogued. Much of this work
was being done in Cambridge and was listed in the
Third Cambridge Catalogue. Although a great deal
of new information was being accumulated from
these radio sources, it was almost impossible to do
proper optical verification because of the extreme
remoteness of the sources. At last the astronomers
got lucky. A strong radio source, a quasar known as
Quasar 3C 273, was occulted, or covered up, by the
moon and this enabled its position to be plotted
very accurately. It appeared at first to be a faint blue

star, but then Maarten Schmidt, an astronomer at Palomar, made a dramatic discovery when he examined the optical spectrum. The four spectral lines which distinguish hydrogen from all the other elements in the universe were shifting to much longer wavelengths, and these red shifts indicated that not only were these quasars very far away indeed, but that they were tremendously powerful. When the light from a source is red shifted, it means that the source is receding and the dramatic red shift in 3C 273 means that the speed is tremendous. It is reckoned that this quasar is travelling at a speed of around 30,000 miles per second and is at present some 2 billion light-years distant from us.

Hundreds of quasars have now been identified and they appear to be the nuclei of active galaxies. They are small, probably only light-days in diameter, but are very luminous, the most luminous objects in the universe. They are possibly powered by immense black holes and it may be that a quasar is a stage in the evolution of a galaxy.

It is now possible to study the positions and make-ups of the galaxies in space, but there is undoubtedly a huge amount of material in space which we are unable to see. It seems certain that the same is true of our own galaxy. The galaxy is rotating and the separate stars are orbiting the centre, but the stars closest to the centre are not orbiting faster than the farther out stars, as they should do according to Kepler's Laws. This points to the probability that the mass of the galaxy is not concentrated in the centre, which then suggests that there is a great deal of unseen matter. Some calculations infer that everything which can be seen in the universe amounts to no more than 10 per cent of the total mass of the universe. The remaining 90 per cent has not yet been detected and perhaps cannot be detected. The question of the missing mass is one of great importance but one to which, at present, there is no answer. Without this answer it is not possible to solve the puzzles of whether the present expansion of the universe will continue, whether the universe will come to an end, or how the universe came to be created in the first place.

Although astronomy has produced a remarkable amount of information and knowledge since the days when man believed that the earth was the centre of the universe, there are still more questions than answers, and many of these questions are fundamental. There are two theories regarding the birth of the universe, the steady state theory or the big bang theory. The steady state theory states that the universe has always existed and that its past is infinite. The big bang theory states that the universe came into being at a specific point in time, on present reckoning about 15,000 million years ago. These theories and the questions related to them will not be resolved until more concrete facts are known, which means that astronomers will continue to be in employment for the foreseeable future. The advent of space travel and exploration has opened a whole new chapter and perhaps the refurbishment of the Hubble Space Telescope will allow man to see further and further, perhaps even to the outer limit of the universe.

CHAPTER 15
The Heavens – A Viewer's Guide

Hopefully this book will have helped to inform and stimulate interest in the amazing worlds around us. The scientific details of the composition and motion of the planets, stars, asteroids and the many other denizens of the universe, have been kept to a minimum. The reasons for this are twofold. Firstly, this book is an introduction to the subject of astronomy and it is important not to frighten off the readers with a non-scientific background. It is certainly not essential to be a scientist to enjoy and be thrilled by this ancient subject and there are many books available for those who wish to delve more deeply into the scientific aspects. The second reason is that the scope of this book does not allow sufficient space to cover every facet of the subject.

For the same reasons it is not possible to provide elaborate details and charts of the stars. Here again, there are other publications which devote themselves to mapping the skies, and it is hoped that those wishing to take their interest further will be encouraged to consult them.

When a cloudless, star-lit sky presents itself, and the weather is not too cold, there is nothing more wonderful than raising one's eyes to the heavens and contemplating the beauty and vastness which is always there. In this world where it is usual to receive nothing for nothing, what a pleasure it is to have access to such a brilliant free show! The planets are constant travellers across the sky, and are to be found in the area of the eliptic, which is the apparent path of the sun across the sky, against the background stars of the celestial sphere. This is the path which contains the constellations of the Zodiac, which extends to seven or eight degrees on either side of the eliptic. There are thousands of stars which can be seen with the naked eye and it may appear to be a big task to learn and become familiar with the celestial layout, but it is not as difficult as it first seems. This is where the constellations are so important. We have recognised earlier that the stars grouped together in a constellation really have no relationship with one another and, in fact, are widely separated, often by many light-years. They are only viewed as groups because they are in the line of vision from earth. Since ancient times, however, they have been known and used as signposts to help man in finding his way around the sky.

It is possibly helpful to start at a familiar point and nothing could be more familiar in the sky than Ursa Major, better known as the Plough or, in America, as the Big Dipper. Although the Plough is not the brightest constellation in the sky, it has an easily identifiable shape and is to be found in an area which is not too heavily populated with stars. For centuries seafarers and other travellers have used the Plough as an important navigational aid, because the two end stars point the way to Polaris, the Pole Star, which is only about one degree off the north celestial pole. There are a total of 88 constellations, with 28 in the northern sky, 48 in the southern sky, and the remaining 12 in the area of the eliptic, which was mentioned above. These 12 constellations form the zodiac, so important to astrologers. Many of the southern constellations can be seen at certain times in the northern hemisphere and vice versa. Astronomers no longer use the constellations to plot objects and find their way around the sky, using instead a celestial coordinate system or altazimuth coordinates, which are used on telescopes with that type of mount. The amateur , however, does use the constellations, and it is a pleasurable way to surf the skies as the ancients did. Although the science is much more technical now, one suspects that even the most modern astronomer still enjoys recognising the various groupings and using the names which have come down from ancient times.

We will now consider the prominent features and the changes in the sky from season to season in both the northern and southern hemispheres. The diagrams are intended only as a rough guide and a star-map should be used for greater detail.

Northern Hemisphere – Spring

At this time of year the Milky Way is skimming the horizon and there are fewer bright stars in the sky than in any other season. There are three principal stars which are easily seen in the spring, Arcturus, Spica and Regulus. To find Arcturus, start with the Plough (Ursa Major) and follow the handle featuring the three stars Alioth, Mizar and Alkaid. This will bring you to Arcturus, which is a brilliant orange star. Apart from Sirius, Arcturus is the brightest star which can be seen from the British Isles and has a magnitude just above zero. A little further east is the small constellation of Corona Borealis (Northern Crown), which really does resemble a crown with its little semi-circle of stars. It is quite easy to find, although not of the first magnitude. Continue the curve of the Plough through Arcturus and you will come on Spica, inVirgo. This is a white star, a binary, which is about 250 light-years away and, with a magnitude of 1.0 is 2,000 times more

Northern Hemisphere – Spring

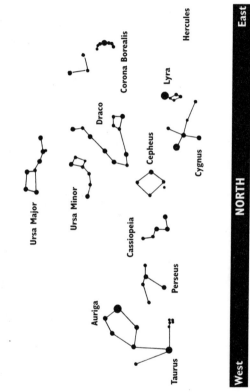

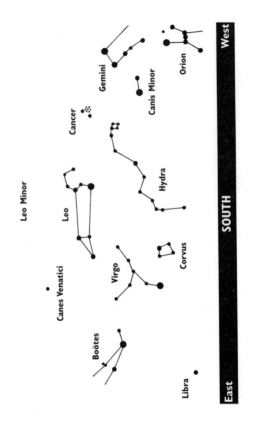

Northern Hemisphere – Spring

luminous than the sun. Leo(the Lion) can be found
by going back to the Plough and by following the
Pointers in the direction away from Polaris. If you
have trouble finding Leo, look for the shape of a
sickle, or a reversed question mark. The brightest
star in the Sickle is Regulus, with a magnitude of
1.3. The Sickle in the middle of Leo is known as an
asterism, which is an arbitrary group of stars with a
recognisable shape within a constellation which
makes the constellation easier to find. One more
interesting star in the spring sky is Alpha Hydrae,
which is also known as the Solitary One, as there are
no other bright stars close to it. This is a reddish
star with a magnitude of 2.0.

Northern Hemisphere – Summer

As we move forward into summer, Orion has gone
from the sky and Ursa Major (the Plough) is in the
north west, with Leo and Virgo low in the west. The
sky at this time has four main bright stars, three of
them forming what has become known as the
Summer Triangle, a name first used by Patrick
Moore. The triangle is made up of Vega, Altair and
Deneb. Vega, in the constellation of Lyra, is almost
overhead at this time and is a lovely blue colour,
which makes it distinctive. Although Lyra is a small
constellation, it does contain some very interesting

Northern Hemisphere – Summer

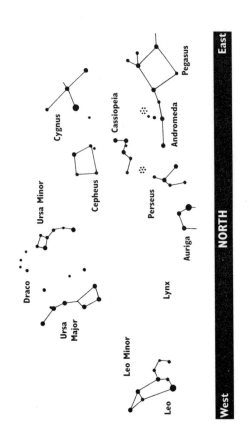

celestial items. There is Epsilon Lyrae, which is a double-double star, Beta Lyrae, an eclipsing variable, and , even more notable, M.57 (the Ring), a planetary nebula. South from Lyra is the constellation of Aquila and there is Altair, the second star in the summer triangle, which can usually be picked out as it lies beween two fainter stars, one orange and one white. Due north from Aquila, slightly east of Lyra, is the constellation of Cygnus (the Swan), high in the sky when looking south. Here is Deneb, the most luminous of the three stars of the triangle. It is reckoned to be about 70,000 times as powerful as the sun and is extremely remote. There are dark patches in the Milky Way in the region of Cygnus and these are caused by black material which is covering the light of the stars beyond the constellation. In the summer skies two constellations of the zodiac are to be found low in the south, Scorpius (the Scorpion) and Sagittarius (the Archer). Scorpius can be recognised by its fishhook-shaped scorpion's tail and also by Antares, the fourth bright star of the summer skies. This a fiery red star with a name meaning 'rival of Mars'. The constellations of Scorpius and Sagittarius are very low in the sky and often are not easily seen in the British Isles. The Milky Way bulges and is very rich with stars in the region of Sagittarius, as it lies in the direction of the centre of our galaxy.

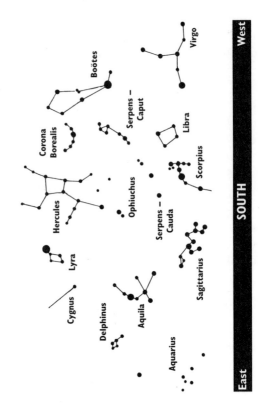

Northern Hemisphere – Summer

Northern Hemisphere – Autumn

The main constellation in the autumn is Pegasus. It is not easy to recognise the form of a horse from the assembled stars, but an asterism called the Great

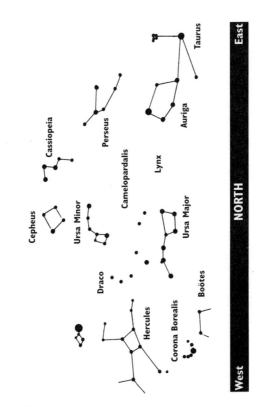

Northern Hemisphere – Autumn

Northern Hemisphere – Autumn

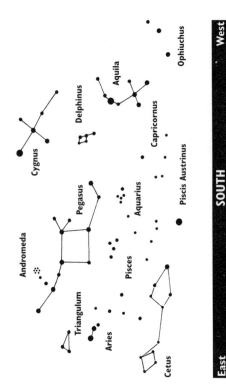

Square, made up of four stars, is easy to find and is situated in the middle of the constellation. These four stars are fairly bright and the only other comparable star in Pegasus is Epsilon, which lies west of the square. The magnitudes of the five stars vary between 2.1 and 2.8. The two stars to the eastern side of the square, Scheat and Alpha Pegasi, point down to a white star called Fomalhaut, with a magnitude of 1.2. It is only 22 light-years away from the sun and is 13 times as luminous as the sun. The evening skies are less bright in autumn than at other times of the year . Scorpius and Sagittarius, which were prominent in the summer, have gone from the sky in autumn and Orion has yet to appear. There are, however, three large constellations of the zodiac to be seen at this time, albeit rather faintly. Below Pegasus lies Pisces (the Fishes), which forms a long line of faint stars, and further south, is to be found Aquarius (the Water-bearer) and Capricornus (the Sea Goat) roughly between Pegasus and the star Fomalhaut. East and slightly south of the Great Square of Pagasus is Aries (the Ram), which can be identified by its two fairly bright stars, which are only 5 degrees apart. North of Aries and east of Pegasus is the galaxy Andromeda, which can best be seen in the autumn and is an amazing 2 million light-years away.

Northern Hemisphere – Winter

The skies on winter evenings in the northern
hemisphere offer the best opportunities to see the
brightest stars. To begin with, Ursa Major and
Orion are both in highly visible positions, the
Plough in the north east and Orion high in the sky
to the south. Orion disappears over the summer
months, but Ursa Major is always on view in all
parts of the British Isles. Follow the pointers in the
Plough and locate the Pole Star, which is in the
constellation of Ursa Minor, or the Little Bear. There
is a similarity between the two Ursas and, of course,
the Plough also has a third name, the Great Bear.
Continue on the line from the two pointers past the
Pole Star and you will come to Cassiopeia, which
has five main stars making a pattern similar to a W
or an M. The three leading stars are about the same
magnitude and one of them, Shedir or Alpha, is
reddish in colour.

Another constellation which is easily recognisable
is Orion, which will be high to the south. The
dominant feature is known as the Hunter's Belt and
is made up of three stars, Epsilon or Alnilam, Zeta
or Alnitak, and Delta or Mintaka. There are two
very bright stars, Alpha or Betelgeux, with a
magnitude of 0.6 and Beta or Rigel, with a
magnitude of 0.1. It is important, from time to
time, to recall the distances involved as we observe
these stars and constellations. For instance, Rigel is

Northern Hemisphere – Winter

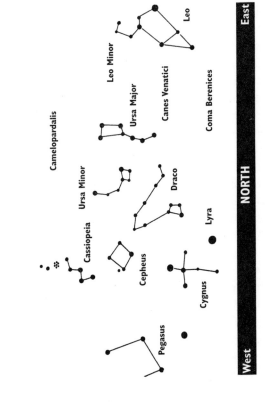

East

Leo

Leo Minor

Canes Venatici

Ursa Major

Coma Berenices

Camelopardalis

Ursa Minor

Draco

NORTH

Cassiopeia

Lyra

Cepheus

Cygnus

Pegasus

West

around 900 light-years away and produces an amazing amount of light, at least 60,000 times as much as the sun. Move below the belt of Orion and you will find a misty cloud , which is of great interest, the Great Nebula. This is a vast cloud of interstellar material which is producing new stars. Although it is over 1,000 light-years away, it can be seen because the side facing the earth is lit by a multiple star called Theta Orionis. The Great Nebula is listed as number 42 in the catalogue compiled by Charles Messier in 1781. Following Orion's Belt down to the south east, the large constellation of Canis Major is easily found. This constellation has several bright stars but they are led by the most brilliant star in the sky, Sirius, which has a magnitude of -1.5. Sirius is a pure white star which appears to twinkle with many colours and this is caused by its position in the sky. Viewed from the British Isles, Sirius is always low in the sky which means that it is seen through the earth's atmosphere, and this disturbs the light and produces the twinkling effect. The higher a star is above the horizon, the less it will twinkle. Orion is a most useful pointer to other interesting stars. To the east is Canis Minor, with its dominant star Procyon, which has a magnitude of 0.4, and to the north east is Gemini, another constellation of the zodiac. Gemini is so called because of two stars known as the Twins, Castor and Pollux. These are easily recognised, Castor being white and a binary, and Pollux being orange. The Milky Way crosses Gemini and lines of stars can be seen stretching from the Twins towards Betelgeux in Orion. Another feature to be seen in this area is an open star-cluster, M.35 in Messier's catalogue, which can be found just south of the Twins. Following the belt of Orion in the opposite direction from Canis Minor, you will come on Taurus (the Bull), with its major star Aldebaran, the Eye of the Bull, which has a magnitude of 0.8. Moving on in the same direction away from Orion lies one of the very well known objects in the sky, the Pleiades. Also known as the Seven Sisters, this is an open cluster which is easily seen with the naked eye and was used in their calendars by the ancients. Two constellations can be found high in the sky during the winter months, Auriga (the Charioteer) and Perseus. Auriga features the bright star Capella, with a magnitude of 0.1 and yellowish in colour. Perseus, to the west of Auriga, has many stars, of which the most interesting is Beta or Algol which varies between 2 and 3 in magnitude every two and a half days. This is an example of an eclipsing binary, in which one partner moves in front of the other, hiding it for a short time. Algol is also known as the Demon Star and appears to give the earth a slow wink. The constellation of Andromeda is to be found between Perseus and the Square of Pegasus, which we have already mentioned. It is in this constellation that is found the most famous of the outer galaxies, the Great Spiral. It can easily be seen with binoculars and is faintly visible with the naked eye, even though it is 2,200,000 light-years away. This is a spiral galaxy like our own but is larger and, therefore, has more than our galaxy's number of 100,000 million stars.

We will turn now to consider the heavens in the Southern Hemisphere and the points of interest to look out for from season to season. One big difference between the hemispheres is that the southern hemisphere does not have the focal point of a bright polar star, as in the north. The southern polar region is not rich in stars and one of the best points of reference is Orion when it is visible, which is not the case during the winter. Ursa Major is not often to be seen in the southern sky, and then usually close to the horizon.

The best known group of stars in the southern sky is the Crux Australis, or the Southern Cross. This is the smallest constellation in the skies, north or south, and it appears on the flags of both Australia and New Zealand. It is interesting to note that Australia uses five stars to represent the Crux, while New Zealand uses only the four major stars. The Crux is to be found surrounded by the constellation Centaurus (the Centaur) and the main stars are Acrux or Alpha Crucis (0.8), Beta (1.2), Gamma (1.6) and Delta (2.8). Three of these stars

Northern Hemisphere – Winter

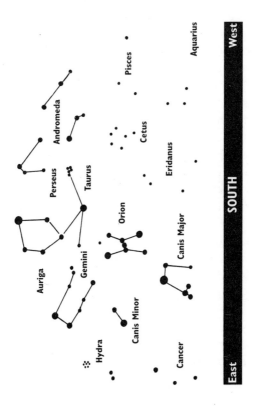

are white and hot and the fourth, Gamma, is a red giant. Until 1679 the Southern Cross was considered part of Centaurus and two of the stars in the constellation are known as the Pointers as they show the way to the Cross. These stars are Alpha Centauri, with a magnitude of -0.3 and Beta (0.6) and, although they look like neighbours, Alpha Centaui is little more than 4 light-years away, compared with Beta's 450 light-years. Beta is a huge star which is more than 10,000 times as luminous as the sun. Next to Sirius and Canopus, Alpha Centauri is the brightest star in the sky and is the nearest bright star beyond the sun. On the other side of the South Pole from the Southern Cross is the bright star Achernar, with a magnitude of 0.5. A good way to find the region of the pole is to look between the Cross and Achernar. The stars in the whole region are dim and the situation is not improved by mist and fog.

Southern Hemisphere – Spring

Many of the stars and constellations which were described in the survey of the northern skies make their appearance again in different seasons in the southern skies. We will mark their positions in the sky but will not repeat the descriptions of individual stars. In October in the southern hemisphere, Orion

Southern Hemisphere – Spring

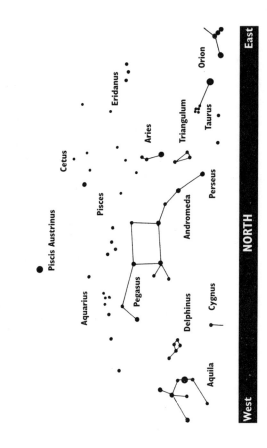

Southern Hemisphere – Spring

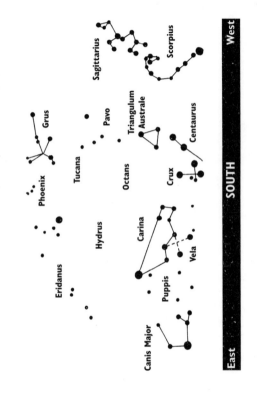

is coming into view low in the east and the Southern Cross is very low in the south,with the bright star Achernar almost overhead. The other major constellation which can easily be found is Pegasus in the north, with the line of stars leading off to Andromeda.

Southern Hemisphere – Summer

In many ways, what is seen in the southern sky is a mirror image of what is seen in the northern sky. In the summer Orion is high in the sky in the north, although the Great Nebula is on the other side of the belt, above instead of below. Sirius is high and does not twinkle as much as it does when seen from the British Isles because it is clear of much of the atmosphere which changes the steady light. The Southern Cross is at this time rising in the south east and the Pointers are also making their appearance. North of the Cross is Carina, with its lead star Canopus, which has a magnitude of -0.7. This whole area is full of nebulae and clusters and bright stars, in addition to a bright display from the Milky Way. In Carina there is a formation of four stars which could be mistaken for the Southern Cross. It is known as the False Cross and the magnitudes of the four stars range from 1.9 to 2.5. Another similarity with the Southern Cross is the fact that

Southern Hemisphere – Summer

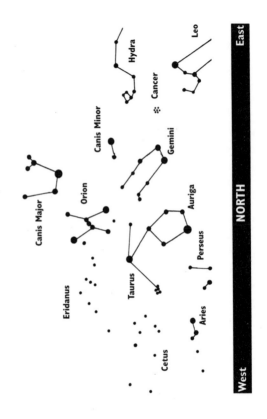

Southern Hemisphere – Summer

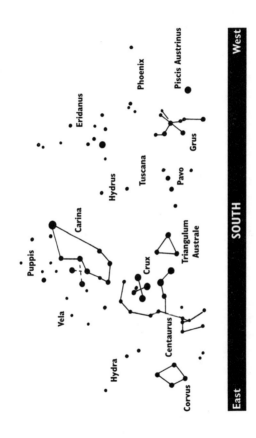

there are three white stars and one orange. Not far from Canopus is a small constellation called Pictor (the Painter) which is interesting as it contains a star, Beta Pictoris, which, it is thought, could be the centre of a system of planets. At this time, the large constellation of Hydra is to be seen in the east, and in the northern sky, also in the east, Leo is just rising.

Southern Hemisphere – Autumn

This is the time of year when the Southern Cross and also Centaurus are almost overhead. The other constellations in evidence are Hydra, which is high in the north west, and also Virgo, Leo and Boots. Orion has almost disappeared, although Sirius is still in view. Orion is now replaced by Scorpius, which is prominent in the south east, and it also is a fine sight. High up in the tail of Scorpius is Lambda Scorpii, which has a magnitude of 1.6 and very close is Upsilon Scorpii with 2.7. Although they appear to be a double star, they are very far apart, Upsilon being four times further away than Lambda.

Southern Hemisphere – Autumn

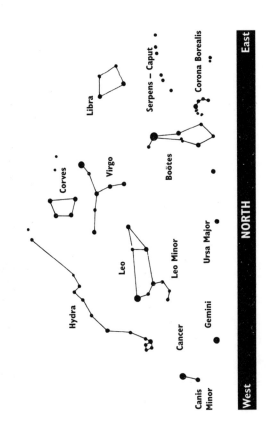

Southern Hemisphere – Autumn

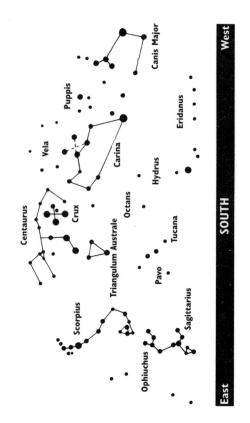

Southern Hemisphere – Winter

Scorpius has now moved into a dominant position and is almost overhead and so also is Sagittarius. The Southern Cross is high in the south west and Canopus has almost left the sky. The four Southern

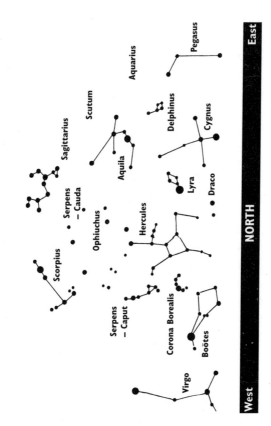

Birds can be seen high to the south east and they are Grus (the Crane), Pavo (the Peacock), Tucana (the Toucan) and the Phoenix. Grus is the most easily seen and its two leading stars are Alpha, white with a magnitude of 1.7, and Beta, which is orange with a magnitude of 2.1. Tucana is not dominant, but it is in the area of the Small Cloud of Magellan and the globular cluster 47 Tucanae. The Cloud is a separate galaxy which is 170,000 light-years away, but the cluster is in our own galaxy. There is also the very bright Large Cloud of Magellan high in the sky. Here is the Tarantula Nebula, which is much larger than M.42 the Great Nebula in Orion.

This short guide to the viewing of the stars has, of necessity, been very basic but, it is hoped, will provide a beginning to the appreciation of the wonderful free display which is given by the heavens. All of the objects which have been mentioned do not appear on the sketch maps and it is strongly recommended that a detailed star-map be used. The weather in the British Isles does not always provide the opportunity to star-gaze, but a guide to what can be seen during the various seasons can help to focus more quickly on what is on view at a particular time.

Southern Hemisphere – Winter

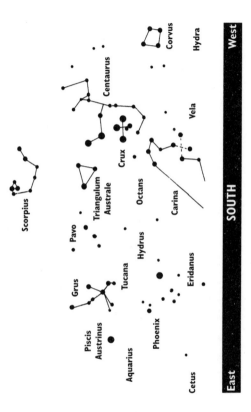

CHAPTER 16
The Continuing Quest

Looked at in relation to the age of the universe, astronomy is only in its infancy. The modern science really began with the invention of the telescope 400 years ago, and at the beginning of the 21st century new discoveries and theories are coming thick and fast. During the writing of this book, there have been constant reminders of the work which is taking place all over the world, and each advance and theory gains the attention of the media. The following are short extracts from stories carried by newspapers, radio and television in recent days.

"A hi-tech telescope has given astronomers a unique and dramatic glimpse into the core of our galaxy, the Milky Way. The Gemini North telescope, sited on Hawaii's Mauna Kea, has revealed clearly for the first time a fast-moving star ploughing through a gas and dust cloud near the centre of the galaxy. Before now, the object, known as IRS8, was merely an ill-defined smudge. Passing through the cloud, the star creates a bow-shock wave, similar to that which forms in front of a boat. The discovery could alter scientist's understanding about the nucleus of the Milky Way, which is believed to harbour a massive black hole...Gemini North is the first of two near-identical eight-metre telescopes, one in the northern hemisphere and the other, in central Chile, in the south... Images from the telescope are on http://www.gemini.edu/galactic.html."

"A new theory has emerged regarding the composition of Mars. An Australian astronomer thinks that Mars has been almost waterless and that water did not create the canyons, as has been thought until now. He believes that the canyons were caused by carbon dioxide escaping from under the rocks of the surface."

"Space shuttle astronauts finally entered the new international space station yesterday after spending four days working on the outside of the craft. The Discovery crew took in supplies, tested air samples and left a message for the station's first residents, scheduled to move in early next month. The shuttle will undock from the station later today... Meanwhile, earth-like clouds and heavy showers produce depressingly familiar weather on Saturn's giant moon, Titan, a new study claims. But on Titan it rains methane instead of water. Astronomers, using the Gemini Observatory in Hawaii, analysed infra-red light emitted by Titan and found variations that indicated the presence of small cloud formations. The clouds dissipate on a daily basis, suggesting that they are showering rain on to the surface. Rain on Titan consists of enormous raindrops falling at a snowflake-like pace through the thick atmosphere, the research suggests".

"A runaway star just 12 miles across but with the mass of the sun has been spotted streaking across the galaxy. The unusual object, a neutron star officially designated as RXJ185635-3754, was captured by Nasa's Hubble space telescope just 200 light-years away in the southern constellation Corona Australis, making it the closest one to earth ever found. Scientists believe it will pass our solar system in about 300,000 years – but at the safe distance of about 170 light-years. It was created in a supernova explosion that would have been visible from earth one million years ago. Scientists believe it has the same mass as the sun packed into an area about 12 miles in diameter."

"A medical student could become the first Scot in space after being awarded a prestigious research placement with the US agency, Nasa... The project's management team have asked her to return to the programme, which is part of a long-term project which aims to put a human on Mars... If she realises her ambition to become an astronaut, her role would be to look after the health of her colleagues during the calculated two year trip to Mars".

These short extracts illustrate the fact that astronomy and space travel figure prominently in our lives today. The space shuttles are launched regularly by the Americans and are making possible the carrying out of scientific experiments of all kinds, as well as increasing experience in space travel. The new international space station, which is a joint venture between the Americans and the Russians, will soon be operational. Unmanned space probes will continue to expand our knowledge deeper and deeper into space. All of this endeavour is supported by the constant work being carried out by earthbound astronomers, both professional and amateur. Astronomy may be in it's infancy, but it is a very lusty infant!

One of the great mysteries for man is what, if anything, is 'out there'? Theories abound about life in outer space, and we have all enjoyed the possibilities and the fantasies which are acted out in books, cinema and television. Unidentified flying objects, or UFO's, are regularly reported in the skies and even on the ground. Strangely, there appears to be a great deal of this type of activity in West Lothian in Scotland! Very many people throughout the world do believe that there is life 'out there', but proof is certainly very thin on the ground. In spite of the concentrated investigations and observations carried out by huge and extremely powerful telescopes on earth and in space, the space flights, the unmanned probes, and the efforts of radio astronomy, not a flicker of life outside of earth has been found. It is incredible that across the immense distances with which we have tried to grapple in this book, the only signs of life would seem to be on this planet of ours. The earth is the fertile jewel in a spinning, barren universe. The present most accepted theory on the birth of the universe is the big bang theory. Could this be a theory which fits in with, and perhaps even confirms, the explanation of the beginning of the world given in the bible? Certainly, many astronauts have spoken of the emotional and religious feelings they have experienced during their missions.

The quest for greater and greater knowledge of our surroundings, our galaxy, the galaxies beyond and the secrets of deep space will continue, no doubt aided more and more by the remarkable advances in technology. In the midst of all this scientific endeavour, however, we must never forget to look up and enjoy the splendours of the heavens.

Absolute magnitude
The apparent magnitude of a star if it could be
viewed from a distance of a fixed 32.6 light-years.
This is a more exact expression of a star's luminosity
than apparent magnitude, which is affected by
distance.

Accretion
The building up of the mass of one body from
another by the mutual gravitational attraction of
matter.

Albedo
The power of a planet or other celestial body to
reflect light.

Angular momentum
The tendency of a rotating object to continue
rotating unless the object is influenced by an outside
force.

Angular separation
The distance between celestial bodies expressed as an angle instead of a linear measurement.

Aperture
The diameter of the main lens or mirror of a telescope, or the main lenses of a set of binoculars.

Aphelion
The point in an object's orbit when it is furthest from the sun.

Apogee
The point in their orbits when the moon, artificial satellite or other object is furthest from the earth.

Apollo asteroids
Asteroids with orbits which cross the orbital paths of the earth and other planets.

Apparent magnitude
The apparent brightness of a star or other celestial body. This brightness is expressed as a number of magnitude and the lower the number, the greater the magnitude.

Asterism
A grouping of apparently related stars in a constellation which has an easily recognised shape

and can be used as a guide to the constellation. An example is Orion's Belt.

Asteroid
A small rocky body in the solar system which orbits the sun. The bigger asteroids are sometimes known as planetoids or minor planets.

Astronomical unit
A unit of measurement which is equal to the mean distance between the earth and the sun, 92,975,000 miles.

Aurora
Electrified particles from the sun rising into the upper air produce these polar lights, which are known in the northern hemisphere as aurora borealis and in the southern hemisphere as aurora australis.

Azimuth
An angular measure in the sky beginning with 0 degrees in the north and then through east, south and west.

Barycentre
The centre of gravity of the system containing the earth and the moon, with the centre being located well within the sphere of the earth.

Big bang
One of the theories regarding the birth of the universe. The theory advocates that an explosion began the process and led to the expansion of the universe.

Binary star
A system of stars in which two stars are physically associated and orbit one another.

Black dwarf
A star which has become completely burned out.

Black hole
This is the result of the collapse of the core of a very large star, in which the area surrounding the star becomes so dense that not even light can escape from it.

Brown dwarf
A star which has failed because the temperature of the core has not risen enough to set off the required nuclear fusion.

Caldera
A huge volcanic crater.

Cardinal points
These are the directions of due north, south, east

and west.

Cassini division
The dark gap between the two rings of Saturn.

Celestial equator
An imaginary line which separates the celestial sphere into the northern and southern hemispheres.

Celestial sphere
An imaginary sphere surrounding the earth, which was once believed to exist with the stars in place. It is now used by astronomers as an easy way to describe the positions of stars as seen from earth.

Cepheid
A very regular, short period, variable star.

Chromosphere
The part of the sun's atmosphere which lies above the photosphere, or bright surface.

Circumpolar star
A star which is very high in the sky, near the celestial pole, and never sets from a particular latitude.

Coma
The nebulous surroundings of the head of a comet.

Comet
A small body made up mainly of ice and dust which orbits the sun. As it nears the sun, some of its material is vaporised, creating the coma and the long tail.

Conjunction
The apparent close approach of two celestial bodies.

Constellations
These are formations of stars which make shapes as seen from earth. There are a total of 88 official constellations.

Core
The innermost layer or centre of a planet or a star.

Core hydrogen burning
This is the nuclear fusion reaction in a star, which is caused by the hydrogen in the core being fused into helium and releasing vast amounts of energy.

Corona
The top layer of the sun's atmosphere.

Cosmic rays
Atomic particles travelling at great speed, which reach the earth from deep space.

Cosmology
The study of the universe.

Critical density
The density which would make the difference between a universe which is expanding infinitely, and one which will eventually collapse.

Dark matter
The vast amount of material in the universe which cannot be detected or seen.

Declination
The angular distance of a body north or south of the celestial equator.

Dichotomy
The precise half-phase of the moon or an inferior planet.

Differential rotation
The atmospheres of the sun and the outer planets rotate at different speeds at their equators and their poles. This is a property of all rotating objects which are rigid.

Doppler effect
The seeming alteration in wavelength of a light source in relation to the movement of the source.

This can be either approaching or receding from the observer.

Earthshine
The faint light which shows on the area of the moon which is not lit by the sun. This is caused by light reflected onto the moon from the earth.

Eccentric
An eccentric ellipse is one which is not circular.

Eclipse
When one body passes in front of another, blocking the light from the body which has been shadowed, or occulted. A lunar eclipse is when the earth comes between the sun and the moon, and a solar eclipse is when the moon comes between the sun and the earth.

Ecliptic
This is the apparent path of the sun as it seems to travel among the stars throughout the year. It is, of course, the projection of the orbit of the earth on to the celestial sphere.

Electromagnetic radiation
These are all kinds of energy derived from rapidly fluctuating electric and magnetic fields, such as radio, light, infrared, ultraviolet, X-ray, and gamma

ray radiation.

Electromagnetic spectrum
The complete range of electromagnetic radiation.

Elliptical galaxy
A galaxy in which no disk or bulge can be detected and appears to be like a circle of stars.

Escape velocity
The minimum velocity required by a body to escape the gravitational pull of another body, such as a planet. The escape velocity on earth is 7 miles per second.

Event horizon
The area surrounding a black hole within which nothing can be seen by outside observers.

Faculae
Bright areas on the surface of the sun.

Fireball
A spectacular and colourful display in the sky caused by the burning up of a meteoroid in the atmosphere.

Frequency
The measurement of wave crests passing a given

point, usually expressed in hertz.

Galactic bulge
A bulge at the centre of our galaxy which is filled with old stars and extends some thousands of light-years from the centre of the galaxy.

Galactic disk
The thinner part of the galaxy which contains old and new stars, as well as gas and dust. It extends some 50,000 light-years from the centre of the galaxy.

Galaxies
Independent systems of stars.

Galaxy cluster
A group of galaxies which are held together gravitationally.

Geocentric
This is the model which places the earth at the centre of the solar system.

Gibbous phase
The phase of the moon, or a planet, between half phase and full phase.

Globular clusters

Huge clusters of stars which are to be found above and below the galactic disk.

Heliocentric
This describes the sun at the centre of the solar system, with the planets and other bodies orbitting the sun.

Hertzsprung-Russell Diagram
Also known as the H-R diagram, stars are plotted in the diagram according to their spectral types and brightness.

Inferior planets
Mercury and Venus are known as the inferior planets because they are closer to the sun than the earth is.

Infra-red radiation
This is a wavelength of light that is longer than the wavelength of visible light and shorter than that of radio waves.

Interferometer
The name given to radio telescopes which are linked together to increase the resolving power.

Interstellar matter
This relates to the gas and dust which is found throughout space. It is the matter which forms stars

and it amounts to about 5 per cent of the mass of our galaxy.

Irregular galaxy
A galaxy which appears to have little shape, but contains the material necessary for the creation of new stars.

Jovian planets
These are the four outer planets which are farthest from the sun. They are Jupiter, Saturn, Uranus and Neptune.

Kiloparsec
This is a measurement made up of 1,000 parsecs, or 3,260 light-years.

Libration
The slow oscillation of the moon as it orbits the earth. This apparent tilting allows a small part of the far side of the moon to be seen.

Light-year
The distance covered by light in one year, 5.88 million million miles.

Local group
The group of galaxies which includes our own galaxy, Andromeda and several other galaxies.

Luminosity
The energy radiated by a star. Another name for luminosity is absolute brightness.

Lunation
The time between new moons, 29 days, 12 hours and 44 minutes.

Magnetosphere
This is an area above the atmosphere of a planet which is part of the magnetic field of the planet.

Magnitude
A classification system for grading stars according to their apparent brightness. Although this is a wide range of stars, the naked eye is normally able to see from magnitude 1, the brightest, to magnitude 6, the faintest.

Main sequence
Stars are analysed for temperature and luminosity and then plotted on the Herpzsprung-Russell diagram (see above). The great majority of stars fall into a band on the diagram, known as the main sequence.

Maria
This is the plural of the Latin mare, the sea, and these words have been applied to dark plains on the

surface of the moon because observers, before the days of the telescope, thought that they looked like expanses of water.

Meteor
Also known as a shooting star, a meteor is a piece of debris from a comet which can be seen streaking across the sky as it burns up in the earth's atmosphere.

Meteorite
A solid body which survives the atmosphere and lands on earth. These are different from meteors and come from the asteroid belt.

Meteor shower
A display in the sky when the orbit of the earth crosses the debris of a comet.

Micron
One thousandth part of a millimeter.

Minor planet
A small planetary body, also known as an asteroid. Most are found in the area between the orbits of Mars and Jupiter.

Nebula
A cloud of gas and dust in space which shows up as

. a fuzzy patch in the sky. The plural is nebulae.

Neutron star
The dense remnant of a massive star after a supernova explosion. It is the core of the star and is made up of neutrons.

Nodes
The points where the path of the ecliptic, the apparant path of the sun, is crossed by the orbit of the moon, a planet or a comet.

Nova
A star which brightens very suddenly and dramatically, due to some kind of collision with material from its binary partner.

Nuclear fission
A reaction in which energy is released by splitting an atomic unit into fragments.

Nuclear fusion
A reaction in which energy is produced by the joining of atomic nuclei.

Obliquity of the ecliptic
This is the angle between the celestial equator and the ecliptic.

Occultation
One celestial body being covered by another celestial body.

Opposition
This occurs when a planet is exactly opposite the sun.

Orbit
The track of a celestial body around another celestial body.

Orbital period
The time taken by a celestial body to complete an orbit.

Parallax
The seeming movement of a body against its background when seen from several different viewpoints.

Parsec
A distance of 3.26 light-years, which is the distance where a star would have a parallax of one second of arc.

Penumbra
This is the partially shadowed region which surrounds the full shadow during an eclipse.

Perigee
The point when the moon is closest to the earth in its orbit.

Perihelion
The point when a planet or any other celestial body is closest to the sun in its orbit.

Photon
The smallest unit of light.

Photosphere
The bright surface of the sun.

Planetary nebula
This is a confusing term, as it is not a planet! It is a hot, white dwarf star which is surrounded by a cloud of gas.

Planetesimals
These are planets in a very early formative stage, possibly no larger than a small moon. They develop from this stage into protoplanets, and then into mature planets.

Precession
The slow alteration in the direction of the axis of the earth due to external influences.

Prominences

Clouds of hydrogen gas which rise from the bright surface of the sun.

Proper motion

The movement of a star on the celestial sphere which can be measured over long periods of time.

Pulsar

A rapidly rotating neutron star which sends out pulsed radio waves.

Quadrature

The position of the moon, or any other celestial body, when seen from earth to be at right angles to the sun.

Quasar

The very first quasars were found at radio frequences. They are very remote and very bright, and are reckoned to be the nucleus of an active galaxy.

Radar

Radar was developed during the Second World War, but in astronomy it is used to measure distances in the solar system and beyond.

Radial velocity

The coming or going movement of a celestial body in relation to the observer.

Radio galaxy
A galaxy which is recognised by its strong radio emissions.

Radio telescope
This is a large dish antenna which is connected to a receiver and also to either recording or imaging equipment. It is used to detect, analyse and record radio emissions from stars and other celestial bodies.

Radio window
The name given to the times it is possible for radio waves from or to space to penetrate the earth's atmosphere.

Red giant
This is the name for the later stages of the life of a star of a mass similar to our sun. The lower surface temperature results in the red colour.

Redshift
The change in wavelength of a celestial body as it recedes from the observer.

Resolving power
This is the power of a telescope to distinguish

between very remote bodies which are close
together.

Retardation
The difference in the time of the rising of the moon
from one night to the next.

Retrograde motion
An orbit or a rotation which is in the opposite
direction from those of the earth or the planets.

Seyfert galaxies
These are galaxies which resemble spiral galaxies.
They tend to be bright and they emit strong radio
signals.

Sidereal period
The period taken by the earth or other planets to
orbit the sun. It is also taken by a satellite to orbit a
planet.

Sidereal month
The time taken by the moon to orbit once around
the earth, 27.3 days.

Sidereal year
The time taken by the earth to orbit the sun,
365.256 days.

Solar day
A day measured from sunrise to sunrise, noon to noon, or sunset to sunset.

Solar flare
A brilliant and powerful eruption above the surface of the sun.

Solar nebula
The immense cloud of gas and dust from which it is thought that the sun and the solar system were formed in the beginning.

Solar wind
A continuous flow of radiation and particles that are emitted from the sun. This wind can actually be seen as it blows the tail of a comet approaching the sun.

Solstices
These are the periods when the sun is at its maximum declination, 23.5 degrees north or south.

Spectral classification
A star classification system based on measuring the surface temperatures spectrographically.

Spectroscope
An instrument which takes incoming light, passes it

through a slit and prism, and divides it into its component colours.

Spectroscopic binary

A binary star system where the two stars are so close together that they cannot be seen separately from earth. They can, however, be detected by the Doppler shifts in their spectra.

Stellar occultation

This is when a planet moves in front of a star, dimming its light.

Sunspots

Dark areas which appear on the face of the sun. They are of irregular shape and the darkness is caused by the fact that they are cooler than the surrounding material.

Superior planet

A planet with an orbit further from the sun than the orbit of the earth.

Supernova

A colossal explosion which accompanies the collapse of a massive star.

Synchronous orbit

When a body's rotation is the same as its average

orbital period. This happens with the moon, which results in the moon always showing one face only to the earth.

Syzygy
The point in the orbit of the moon when it is new or full. Either of the two positions, conjunction or opposition, of a celestial body when the sun, the earth and the body lie in a straight line.

Terminator
The dividing line between the sunlit and night sides of the moon or a planet.

Terrestrial planets
The four planets in the solar system which are closest to the sun. These are Mercury, Venus, Earth and Mars.

Tidal bulge
The alteration in a celestial body due to the gravitational pull of another body. The moon has this effect on the earth, causing an elongation, or a tidal bulge.

Umbra
The main shadow cast by the earth during an eclipse. The term is also used to describe the darkest part of a sunspot.

Universal recession
The apparent recession of all other galaxies away from our galaxy.

Van Allan Zones
Zones of highly energetic charged particles which surround the earth.

Variable star
A star which changes in brightness over fairly short periods. There are different types of variable star, which are described in chapter 13.

Visual binaries
Binary stars which can be seen with the naked eye or with a telescope.

Wavelength
The distance between two wave crests or troughs, which is measured in metres.

White dwarf
The remnant core of a red giant star, which is very small and dense.

Zenith
An observer's overhead point in the sky.

Zodiac

The belt round the sky which extends to 8 degrees on either side of the ecliptic. This is the area where the sun, moon and main planets can be found.

Zonal flow
The east to west wind system which prevails on Jupiter.

Appendices

The inner planets, those which orbit the sun in our part of the solar system, are the closest neighbours of the earth, making them the most obvious targets for observation and exploration, especially as they can now be reached by modern rockets. In recent years, space probes have also been able to send back more and more information on the very remote outer planets. All of this new knowledge has been greatly augmented by the development of larger and more sophisticated telescopes on earth and the Hubble telescope in space.

The following basic information will be of interest to those who wish to take their study of the planets further.

Planet	Distance from sun in millions of miles	Orbital period	Axial rotation period	Inclination of axis in degrees
Mercury	36	88 days	58.6 days	2
Venus	67	224.7 days	243 days	178
Earth	93	365.2 days	23h 56m	23.4
Mars	141.5	687 days	24h 37m	24
Jupiter	483	11.9 years	9h 50m	3
Saturn	886	29.5 years	10h 14m	26
Uranus	1783	84 years	17h 14m	98
Neptune	2793	164.8 years	16h 7m	29
Pluto	3666	247.7 years	6d 9h	122

Planet	Diameter in miles	Escape velocity miles/second	Gravity Earth=1	Mass Earth=1	Surface temp. C
Mercury	3,030	2.6	0.38	0.06	+427
Venus	7,523	6.4	0.90	0.86	+480
Earth	7,926	7.0	1	1	+22
Mars	4,222	3.2	0.38	0.11	-23
Jupiter	89,424	37	2.64	318	-150
Saturn	74,914	22	1.16	95	-180
Uranus	31,770	14	1.17	15	-214
Neptune	31,410	15	1.2	17	-220
Pluto	1,444	0.7	0.06	0.002	-230

The planets of the solar system have at least 50 known satellites, which differ a great deal in size. The only two planets without satellites are Mercury and Venus, and the Earth has only one, the moon. Many of the satellites have been studied and photographed by the space probes and knowledge on these bodies is constantly being built up. The following basic information gives the size of the satellites and their distances from their primary planets. It is interesting to know when each of these satellites was discovered, and this information has also been given.

Planet and satellite	Diameter in miles	Distance from primary 1,000 miles	Year of discovery
EARTH			
Moon	2,172	240	—
MARS			
Phobos	14	5.86	1877
Deimos	8.125	14.66	1877

Planet and satellite	Diameter in miles	Distance from primary 1,000 miles	Year of discovery
JUPITER			
Metis	25	80	1979
Adrastea	12.5	80.6	1979
Amalthea	119	113	1892
Thebe	62.5	139	1979
Io	2269	264	1610
Europa	1961	419	1610
Ganymede	3288	669	1610
Callisto	3000	1177	1610
Leda	10	6934	1974
Himalia	116	7175	1904
Lysithea	22.5	7325	1938
Elara	47.5	7335	1905
Ananke	18.75	13250	1951
Carme	25	14125	1938
Pasiphae	31.25	14687	1908
Sinope	22.5	14812	1914

Planet and satellite	Diameter in miles	Distance from primary 1,000 miles	Year of discovery
SATURN			
Pan	12.5	84	1991
Atlas	18.75	86	1980
Prometheus	65.5	86	1980
Pandora	56	89	1980
Janus	56	94	1966
Epimetheus	75	94	1966
Mimas	244	116	1789
Enceladus	312	149	1789
Tethys	656	184	1684
Telesto	19	184	1980
Calypso	15.6	184	1980
Dione	700	235	1684
Helene	19	235	1980
Rhea	956	329	1672
Titan	3219	764	1655
Hyperion	181	926	1848
Iapetus	900	2225	1671
Phoebe	138	8095	1898

Planet and satellite	Diameter in miles	Distance from primary 1,000 miles	Year of discovery
URANUS			
Cordelia	16.25	31.25	1986
Ophelia	18.75	33.75	1986
Bianca	26.25	37	1986
Cressida	39	39	1986
Desdemona	34	39	1986
Juliet	52	40	1986
Portia	67	41	1986
Rosalind	34	44	1986
Belinda	41	47	1986
Puck	96	54	1985
Miranda	295	81	1948
Ariel	724	119	1851
Umbriel	732	166	1851
Titania	987	272	1787
Oberon	952	365	1787

Planet and satellite	Diameter in miles	Distance from primary 1,000 miles	Year of discovery
NEPTUNE			
Naiad	34	30	1989
Thalassa	50	31	1989
Despina	94	33	1989
Galatea	100	39	1989
Larissa	125	46	1989
Proteus	262	74	1989
Triton	1687	221	1846
Nereid	212	3444	1949
PLUTO			
Charon	741	12	1978

SELECTED COMETS

Name	Period in years	Distance from sun in millions of miles	
		Min.	Max.
Encke	3.3	32	380
Grigg-Skjellerup	5.1	93	460
D'Arrest	6.2	110	520
Pons-Winnecke	6.3	116	520
Giacobini-Zinner	6.5	92	550
Finlay	6.9	102	580
Faye	7.4	150	560
Tuttle	13.3	95	970
Crommelin	27.9	69	1640
Tempel-Tuttle	32.9	91	1820
Halley	76.1	55	3290
Swift-Tuttle	130	89	4800

METEOR SHOWERS

Name	Maximum Activity	Estimated Hourly Count	Parent Comet
Quadrantid	Jan 3	50	unknown
Lyrids	April 21	10	unknown
Beta Taurid	June 30	25	Encke
Perseids	Aug 12	75	1862III (Swift-Tuttle)
Draconids	Oct 8–9	500+	Giacobini Zimmer
Orionids	Oct 22	25	Halley
Leonids	Nov 17	10	1866I (Tuttle)
Geminids	Dec 13	75	3200 Phaeton

The orbit of the earth crosses the densest part of the debris of the Leonids every 33 years. It is often possible for the estimated count at that time to rise to an incredible 1,000 per minute.

FIRST MAGNITUDE STARS

Name	Magnitude	Declination
Sirius	-1.5	-17
Canopus	-0.7	-53
Alpha Centauri	-0.3	-61
Arcturus	-0.0	+19
Vega	0.0	+39
Capella	0.1	+46
Rigel	0.1	-08
Procyon	0.4	+02
Achernar	0.5	-57
Betelgeux	var	+07
Agena	0.6	-60
Altair	0.8	+09
Acrux	0.8	-63
Aldebaran	0.8	+17
Antares	1.0	-26
Spica	1.0	-11
Pollux	1.1	+28
Fomalhaut	1.2	-30
Deneb	1.2	+45
Beta Crucis	1.2	-60
Regulus	1.3	+12

THE CONSTELLATIONS

Name	Common Name	Hemisphere	Stars of 1st Magnitude
Andromeda	Andromeda	N	—
Antlia	The Air Pump	S	—
Apus	The Bird of Paradise	S	—
Aquarius	The Water-bearer	S	—
Aquila	The Eagle	N	Altair
Ara	The Altar	S	—
Aries	The Ram	N	—
Auriga	The Charioteer	N	Capella
Boots	The Herdsman	N	Arcturus
Caelum	The Graving Tool	S	—
Camelopardalis	The Giraffe	N	—
Cancer	The Crab	N	—
Canes Venatici	The Hunting Dogs	N	—
Canis Major	The Great Dog	N	Sirius
Canis Minor	The Little Dog	N	Procyon
Capricornus	The Sea-Goat	S	—

Name	Common Name	Hemisphere	Stars of 1st Magnitude
Carina	The Keel	S	Canopus
Cassiopeia	Cassiopeia	N	—
Centaurus	The Centaur	S	Agena, Alpha Centauri
Cepheus	Cepheus	N	—
Cetus	The Whale	S	—
Chamaeleon	The Chamaeleon	S	—
Circinus	The Compasses	S	—
Columba	The Dove	S	—
Coma Berenices	Berenice's Hair	N	—
Corona Australis	The Southern Crown	S	—
Corona Borealis	The Nothern Crown	N	—
Corvus	The Crow	S	—
Crater	The Cup	S	—
Crux Australis	The Southern Cross	S	Acrux, Beta Crucis
Cygnus	The Swan	N	Deneb
Delphinus	The Dolphin	N	—
Dorado	The Swordfish	S	—

Name	Common Name	Hemisphere	Stars of 1st Magnitude
Draco	The Dragon	N	—
Equuleus	The Foal	N	—
Eridanus	The River	S	Achernar
Fornax	The Furnace	S	—
Gemini	The Twins	N	Pollux
Grus	The Crane	S	—
Hercules	Hercules	N	—
Horologium	The Clock	S	—
Hydra	The Watersnake	S	—
Hydrus	The Little Snake	S	—
Indus	The Indian	S	—
Lacerta	The Lizard	N	—
Leo	The Lion	N	Regulus
Leo Minor	The Little Lion	N	—
Lepus	The Hare	S	—
Libra	The Balance	S	—
Lupus	The Wolf	S	—
Lynx	The Lynx	N	—
Lyra	The Lyre	N	Vega

Name	Common Name	Hemisphere	Stars of 1st Magnitude
Mensa	The Table	S	—
Microscopium	The Microscope	S	—
Monoceros	The Unicorn	Equat	—
Musca Australis	The Southern Fly	S	—
Norma	The Rule	S	—
Octans	The Octant	S	—
Ophiuchus	The Serpent-bearer	Equat	—
Orion	Orion	Equat	Rigel, Betelgeux
Pavo	The Peacock	S	—
Pegasus	The Flying Horse	N	—
Perseus	Perseus	N	—
Phoenix	The Phoenix	S	—
Pictor	The Painter	S	—
Pisces	The Fishes	N	—
Piscis Australis	The Southern Fish	S	Formalhaut
Puppis	The Poop	S	—
Pyxis	The Mariner's Compass	S	---
Reticulum	The Net	S	—

Name	Common Name	Hemisphere	Stars of 1st Magnitude
Sagitta	The Arrow	N	—
Sagittarius	The Archer	S	—
Scorpius	The Scorpion	S	Antares
Sculptor	The Sculptor	S	—
Scutum	The Shield	S	—
Serpens	The Serpent	N	—
Sextans	The Sextant	S	—
Taurus	The Bull	N	Aldebaran
Telescopium	The Telescope	S	—
Triangulum	The Triangle	N	—
Triangulum Australe	The Southern Triangle	S	—
Tucana	The Toucan	S	—
Ursa Major	The Great Bear	N	—
Ursa Minor	The Little Bear	N	—
Vela	The Sails	S	—
Virgo	The Virgin	Equat.	Spica
Volans	The Flying Fish	S	—
Vulpecula	The Fox	N	—

FORTHCOMING ECLIPSES

Total Solar Eclipses

Where Visible	Date	Duration in Minutes
Southern Africa	2001 June 21	4.9
Southern Africa and Australia	2002 Dec 4	2.1
Antarctica	2003 Nov 23	2
South Pacific	2005 April 8	0.7
Africa, Russia and Near East	2006 March 29	4.1
Arctic, Siberia and China	2008 Aug 1	2.4
South Pacific, India and China	2009 July 22	6.6
South Pacific	2010 July 11	5.3
South Pacific and N Australia	2012 Nov 13	4
Central Africa and Atlantic Oc	2013 Nov 3	1.7

Where Visible	Date	Duration in Minutes
North Atlantic and Arctic Ocean	2015 March 20	4.1
Indonesia and Pacific Ocean	2016 March 9	4.5
U.S.A., Pacific and Atlantic Oc	2017 August 21	2

FORTHCOMING ECLIPSES

Total Lunar Eclipses

2003 May 16

2003 November 9

2004 May 4

2004 October 28

2007 March 3

2007 August 28

2008 February 21